PLC Programming using RSLogix 500

Basic Concepts of Ladder Logic Programming!

Book 1

By

Gary D. Anderson

ISBN-13: 978-1511770347

ISBN-10: 1511770341

Contents

Preface: Goals and Focus!

If you need to gain familiarity and confidence with PLC programming, but have never had the benefit of formal training in programming PLC's, then I believe this book will greatly benefit you!

The best way to gain proficiency in any difficult subject is to develop a solid understanding of the "basics", which then become a foundation for necessary skills and abilities in any field of endeavor. As you can tell by simply reading over the "*Table of Contents*", the whole focus of this book is to take you in a step-by-step journey through "basic" PLC programming concepts. The goal is to gain proficiency and build confidence in your programming abilities.

The use of PLC's and Process Automation Controllers is widespread throughout industry, and the need for technicians who are skilled in these areas, together with electrical and electronics troubleshooting skills, I believe has never been greater than it is today. While the use of PLC's in automation and control is widespread, it is not necessarily always easy or straightforward to program and troubleshoot equipment. There are many different controllers on the market, each with their own unique application and development software, different communication software and protocols, and nuances in their program instruction sets. For these reasons, it is important to develop a solid knowledge of the whole concept of PLC programming instructions, what they do, how they work, and how to build a program from these instructions that will accomplish a desired result.

In the mid 70's I began my journey into the world of troubleshooting machine technology and automation. During that period, this world was pretty much dominated by large grids of electro-mechanical relays and timers that were used to provide the control "logic" these machines required to perform their task. It was not uncommon to see electrical schematics on large 3'x 6' sheets which often made troubleshooting a daunting task. In addition to this, the "logic" of the machine was basically "hard-wired" – control relay to control relay, by what seemed like miles of wire. This made the basic control of equipment very difficult to alter or modify. During the early 80's the company I worked for began integrating PLC's – "***programmable logic controllers***" into existing equipment, and a new world of possibilities for machine and process control opened up. As old machines were replaced with newer PLC controlled equipment, machine downtime became less frequent and we discovered that troubleshooting a fault or problem could be much easier to pinpoint and correct.

Programmable Logic Controllers or "PLC's" were developed during the 1960's to meet the ever-changing needs of the automobile industry, where products changed yearly and made frequent modifications to wiring and relay-based control systems a necessity as well. The ability to modify controls within the "ladder logic" of a machine's "controller", rather than its hard-wired components, made these changes easier for the technician and also more cost efficient for the industry. While relays, both electro-mechanical and solid state, continue to play a major role in controlling machines and process equipment, they are most efficiently used as discrete components in conjunction with some form of programmable logic control.

Of course today, this technology is common to not only the automobile industry, but in petroleum and chemical plants, CNC machining, manufacturing, and in just about every type of industry you can imagine. PLC technology and the ability to program and effectively troubleshoot these types of systems has become a necessary and worthwhile pursuit for anyone involved as a technician or engineer within modern industry.

Today there are many brands of these controllers available on the market: Allen-Bradley, GE Fanuc, Omron, Direct Logic, and Sq-D to name only a few. In terms of programming a new project, there is a definite learning curve to each brand of controller. Yet the basic concepts of "ladder logic" instructions, hardware architecture and I/O functionality, has become pretty standard across all brands of PLC. Consider for example the "ladder logic" of a Fanuc controller of a four-axis CNC machine. While differences may exist in scan times or addressing symbols, the basic concepts remain the same. The overall appearance and usefulness in troubleshooting is a common element of all the PLC's that I have encountered. Here is an example of Fanuc control ladder logic in a multi-axis CNC.

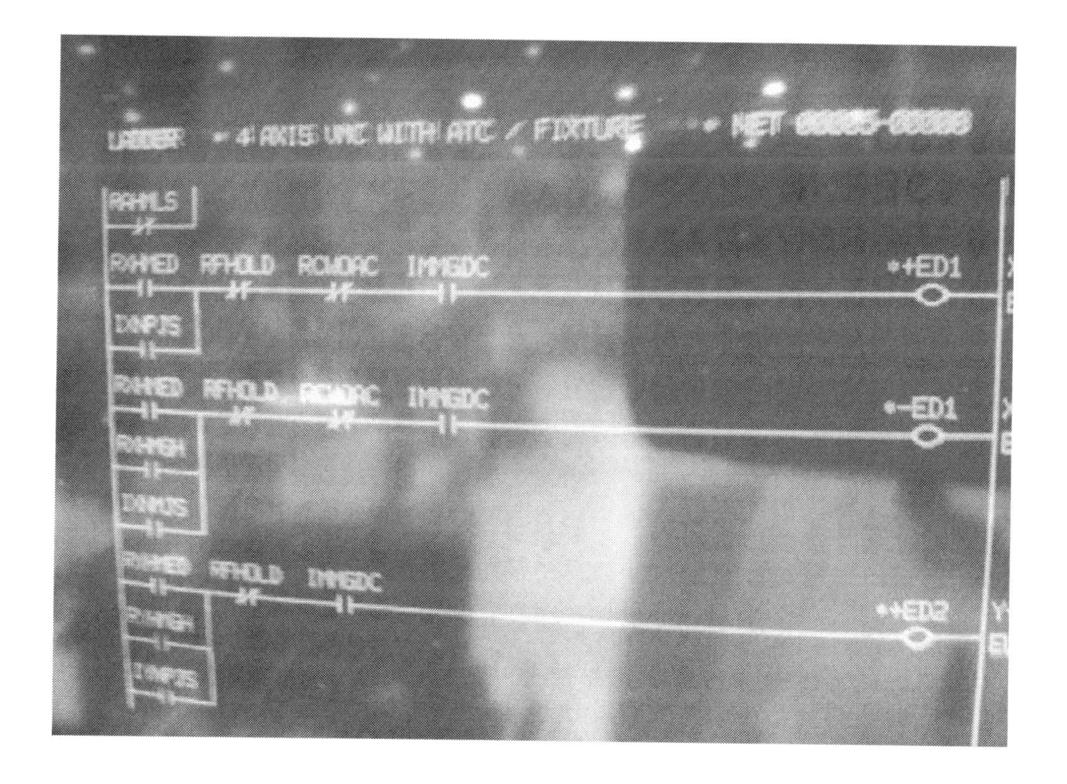

As you can see, the basic structure and elements in this type of program, probably written in C or C+ and burned into the Fanuc controllers EEPROM chips, are essentially the same in function and form as any one of many stand-alone PLC controllers that are on the market.

This three book series, "PLC Programming using RSLogix 500", focuses on PLC concepts and programming using the Allen-Bradley line of PLC's, specifically the SLC500 & MicroLogix controllers and RSLogix500 software. From here it's a reasonably easy transition to other AB controllers with some degree of confidence and understanding, such as the Allen-Bradley PLC 5's, or the ControlLogix and CompactLogix PLCs. Another reason for focus on the RSLogix500 instruction set is due to the wide use and acceptance of Allen Bradley controllers throughout industry, and also because the RSLogix platform is straightforward to use and works well in demonstrating the concepts of "ladder logic" elements and programming. The series is divided into three books:

Basic Concepts of Ladder Logic Programming (Bk1)

Advanced Programming Concepts (Bk2)

Ladder Logic Diagnostics & Troubleshooting (Bk3)

Examples will be shown with clear objectives throughout each section which should, upon completion, enable you to confidently program machinery and processes to accomplish your aims, and certainly to aid you in troubleshooting and repair of machine faults when they occur.

What I'd like to emphasize here is the *practical* nature of this series. It's always great to be in a classroom with an instructor, but with the hectic schedules of many production-driven businesses, it quite often simply doesn't happen. More often than not, "on-the-job" training is the order of the day. If this is where you find yourself, some really practical thoughts, explanations or examples can be a tremendous benefit – especially if production is down and all eyes are upon you to resolve an issue. I hope these books will provide exactly that type of help, and also be a foundation for further learning and exploration of automation and PLC technology.

Best Regards,
Gary D. Anderson

Chapter 1: Introduction to Programmable Logic Controllers!

With the development of PLCs came the capability to employ greater levels of "***automation***" to machine processes and control. Much of the ongoing "monitoring" and "repetition" of machine functions are delegated to ***sensors*** or input devices that provide feedback signals to the PLC processor. The processor or CPU then changes the status of outputs to allow for the changing conditions – all according to the "program" or *ladder logic* you have written and loaded into the PLC.

The PLC can monitor pressures, temperatures, positioning of an axis, liquid level in a tank, flow in a pipeline, when a valve is open or closed, and a myriad of other conditions that at one time required human intervention and oversight. You might say that a programmable logic controller acts as the centralized "brain" of automated machinery and equipment, receiving - much like the human brain, inputs from its sensory devices.

Sensors are the various *input devices* that convert a physical condition to an electrical signal of some form. This signal is then sent to the PLC and enters via the appropriate *input modules*. For this example, even a momentary push-button or maintained E-Stop button, is a type of input device; transmitting perhaps a 120vac signal to an input card within the PLC. Some input devices simply indicate an on/off state (discrete inputs), while other types of inputs transmit a changing status by providing a variable voltage or current to its designated input terminal. An example of this could be a thermocouple or RTD input that provides a constant, though changing, and temperature signal.

Actuators are the various *output devices* that convert electrical signals – both digital and analog, into the different physical conditions the equipment is programmed to perform. This could be increasing a pressure, re-filling a tank by turning on a pump, or enabling

a servomotor to move an axis into its programmed position. These signals are transmitted to output devices from the *output modules* of the PLC, each from a specified terminal.

This chapter's goal, is to build a solid foundational understanding of PLCs and how they function, we start by focusing on the following objectives:

- To better understand the advantages of using a *centralized* controller – a PLC over the use of a strict relay-logic machine control.
- To become acquainted with PLC architecture and hardware components that can be used for your specific needs.
- To understand how input and output devices are connected to a PLC.
- To understand how a PLC interfaces or communicates with application software, such as RSLogix500, to allow online monitoring of a program, downloading new or modified programs in the PLC, or uploading of existing program for maintenance back-up or offline editing.

PLC Architecture & Systems:

The basic architecture common to most PLC's is shown in the following photo. In it you will see, from left to right: a Power Supply, the CPU or "central processing unit", and then a series of other modules that comprise the "input" and "output" or I/O section of the PLC. With the Allen-Bradley SLC500 family, the power supply and CPU always reside in the leftmost slots – the CPU dedicated to slot 0, which is the second from the left. Collectively these connect into a "backplane" which can be specified for the number of slots needed for the number of input or output cards you are using. The input and output cards, or I/O modules, can reside in any order. You will designate which slot they are in when you are configuring a new project.

Here is an example of a four slot SLC 500. As you can see the power supply is far left, the CPU (with the key switch and status LEDs) in the leftmost regular slot and then the next three I/O modules in any order. The right four slots are designated 0, 1, 2, and 3 from left to right. These slot assignments become essential to proper addressing as we write ladder logic programs.

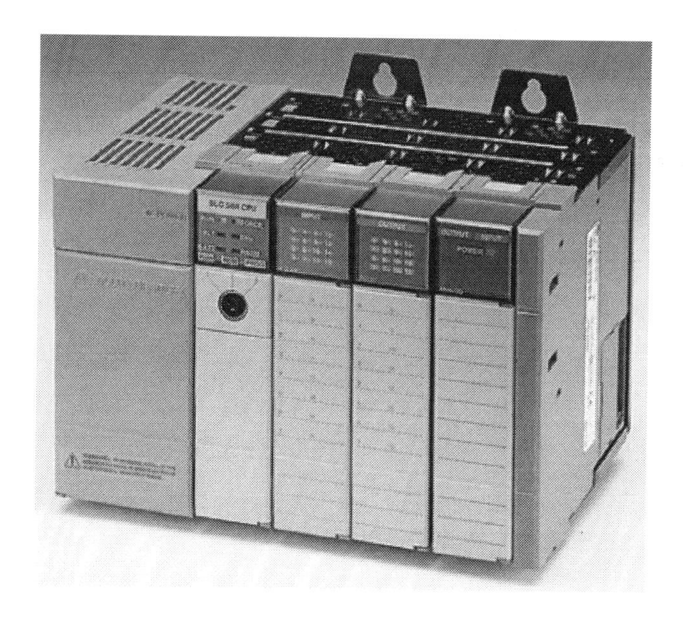

The CPU

The CPU is the main control center where the PLC determines what changes need to be made in the individual outputs, based on the ladder logic programed into its memory. We'll look more into the "scan & update routine" later, but for now just know that the CPU receives input from its sensors and input devices, makes calculations and decisions based upon its program, and then changes the status of its outputs accordingly – every 5 to 50 milliseconds!

Here is an inside look at a SLC500 CPU:

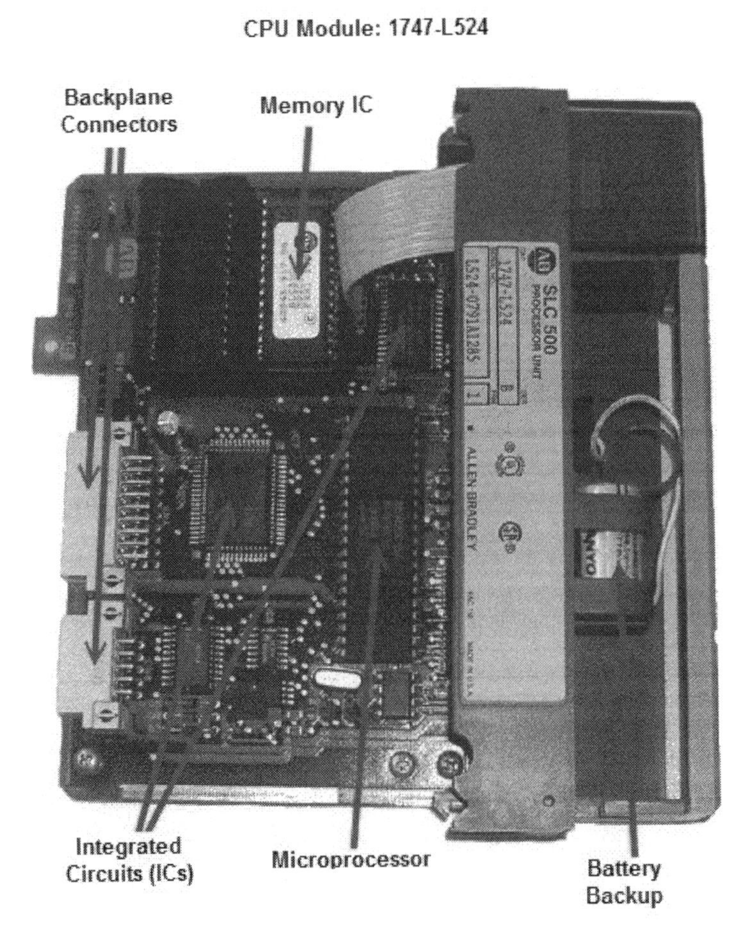

The user or "ladder" program is stored in the RAM or "random access memory" of the CPU. The memory, along with any program that it holds, is maintained as long as power is on or by the battery that is connected to the board. For this reason it is often referred to as "volatile memory". This program is executed in sequential order when the processor is in "RUN" or "REMOTE RUN" modes. If your CPU is equipped with the EEPROM option then your ladder program can also be written to the EEPROM. Upon the event of battery failure and program loss it can be reloaded into the active microprocessor memory.

The CPU also contains ROM or "read-only" memory that contains the processor's firmware program language so that it knows how to respond or act upon your ladder logic. This memory is "non-volatile" and is "burned" into the microprocessors memory. The battery used to maintain the programming in the RAM usually lasts approximately three years and so a good maintenance practice is to employ periodic checks and to record the date backup batteries are placed into service.

Some important diagnostic tools are simply the status indicator LED's on the front of the CPU. They give a quick read into several key processor issues by indicating if the processor is in "RUN" mode, if there is a processor fault, a communications link, the battery is low, and if "forces" are enabled in the ladder program.

The Power Supply

The power supply connects into the backplane of a modular PLC and supplies the power used by the CPU and I/O sections of the PLC. This is electrical power solely for the CPU itself and its communication between with various I/O modules. The numerous devices such as limit switches, push buttons, temperature sensors, solenoids, or relays must be supplied with their own external power source. We will discuss this wiring in more detail in the later chapters.

Programming Interface Device

In most instances, this is the laptop or desktop computer used to write a ladder logic program that has the capability to download, upload or "go online" and monitor an existing program while the machine is operational. In order to do these things our programming device must be able to "communicate" with the desired controller. All PLC manufacturers have their own application and development software that must be used with their PLC. As earlier mentioned, our focus for this series is the SLC 500 family of Allen Bradley PLC's. The application software for programming is the RSLogix500 software from Rockwell Automation.

While this software provides a means of developing or building a ladder program usually an additional communication program is required as well. For the AB PLCs this program is "RSLinx" software which provides the drivers and communication protocols necessary for communication between laptop and PLC.

The following shows an example of typical cable connections between the programming device and the controller CPU.

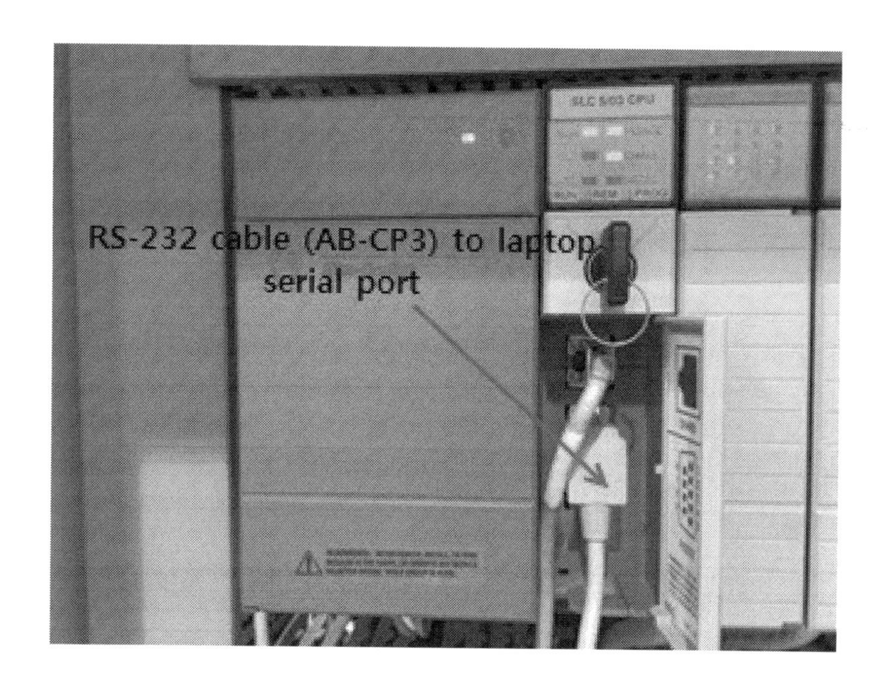

RSLinx provides for a great number of possible types of interfacing and provides drivers and protocols for RS-232, RS-485, DH (Data Highway), DH+ and Ethernet communications. Each different type of communication must be configured for the protocol you need to use which is usually dependent on the model of PLC.

Communication Drivers & Cables:

Here is a chart of the drivers and cables which I routinely use and carry with my laptop. Several of these are USB adapters specifically built for the Allen Bradley protocols, while the CP3 cable is pretty standard for RS-232 serial port connection – perhaps you have an older laptop that still has a serial port. Again these "drivers" can be selected from within the RSLinx software.

Hardware, Adapters and Cable Selection			RS Linx Selections & Settings			
Laptop to the following devices:	**Adapter & Port**	**Cable**	**Driver Types**	**Driver Name**	**Device Type**	**Notes**
SLC 500 Fixed Controller (DH-485)	1747-UIC / RS-485 port	1747-C13	RS-232	AB-DF1-1	1770-KF3/1747-KE	Select Comm Port & CRC Operates at 19.2 Kbps only
SLC 5/01, 5/02, 5/03 Controller (DH-485)	1747-UIC / RS-485 port	1747-C13	RS-232	AB-DF1-1	1770-KF3/1747-KE	Select Comm Port & CRC Operates at 19.2 Kbps only
PanelView 300 or Higher Terminal (DH-485)	1747-UIC / RS-485 port	1747-C13	RS-232	AB-DF1-1	1770-KF3/1747-KE	Select Comm Port & CRC Operates at 19.2 Kbps only
1747-AIC (DH-485)	1747-UIC / RS-485 port	1747-C13	RS-232	AB-DF1-1	1770-KF3/1747-KE	Select Comm Port & CRC Operates at 19.2 Kbps only
PanelView 300 or Higher Terminal (DH-485)	1747-UIC / RS-232 port	2711-NC13	RS-232	AB-DF1-1	1770-KF3/1747-KE	Select Comm Port & CRC Operates at 19.2 Kbps only
SLC 5/03, 5/04, 5/05 Controller (DH-485)	1747-UIC / RS-232 port	1747-CP3, 1756-CP3	RS-232	AB-DF1-1	1770-KF3/1747-KE	Select Comm Port & CRC Operates at 19.2 Kbps only
CompactLogix Controller (DH-485)	1747-UIC / RS-232 port	1747-CP3, 1756-CP3	RS-232	AB-DF1-1	1770-KF3/1747-KE	Select Comm Port & CRC Operates at 19.2 Kbps only
FlexLogix Controller (DH-485)	1747-UIC / RS-232 port	1747-CP3, 1756-CP3	RS-232	AB-DF1-1	1770-KF3/1747-KE	Select Comm Port & CRC Operates at 19.2 Kbps only
ControlLogix Controller (DH-485)	1747-UIC / RS-232 port	1747-CP3, 1756-CP3	RS-232	AB-DF1-1	1770-KF3/1747-KE	Select Comm Port & CRC Operates at 19.2 Kbps only
MicroLogix Controller (DH-485)	1747-UIC / RS-232 port	1761-CBL-PM02	RS-232	AB-DF1-1	1770-KF3/1747-KE	Select Comm Port & CRC Operates at 19.2 Kbps only
AIC+ (DH-485 protocol)	1747-UIC / RS-232 port	1747-CP3, 1756-CP3 (port 1) 1761-CBL-PM02 (port 2)	RS-232	AB-DF1-1	1770-KF3/1747-KE	Select Comm Port & CRC Operates at 19.2 Kbps only
SLC 5/04 and PLC 5 controllers using DH+ protocol (exceptions PLC-5/10, 5/12, 5/15, 5/25)	1784-U2DHP	8 pin DIN / integral to adapter	RS-232	DF1	1770-KF2/1785-KE/SCANport	Select Comm Port & CRC (For RSLinx, Ver 2.52 and earlier)
	1784-U2DHP	8 pin DIN / integral to adapter	1784-U2DHP for DH+ devices	*Enter name or use default name*		Select Comm Port & Station No.(For RSLinx, Ver 2.53 and later)
ControlLogix / CompactLogix / PanelView Plus, etc.	Straight-through cable / to Switch Cross-over cable peer-to-peer		Ethernet or EtherNet/IP			Configure Laptop IP address for the network being connected.

PLC – Input & Output Modules

The input and output modules are, in some respects, the most important components of the PLC. Without the capability to *isolate* and *translate* the "sensory" inputs and then re-translate and send out the appropriate output signal it would be impossible to provide machine control. These critical functions are performed by the input and output modules. Generally the I/O modules deal with voltages and current that are higher than what the CPU can safely process and so provide the necessary "*isolation*" between the input signals being received, the output signals going out, and the CPU and backplane circuitry. Here is a "simplified" block diagram of CPU and I/O relationship.

PLC Section Diagram

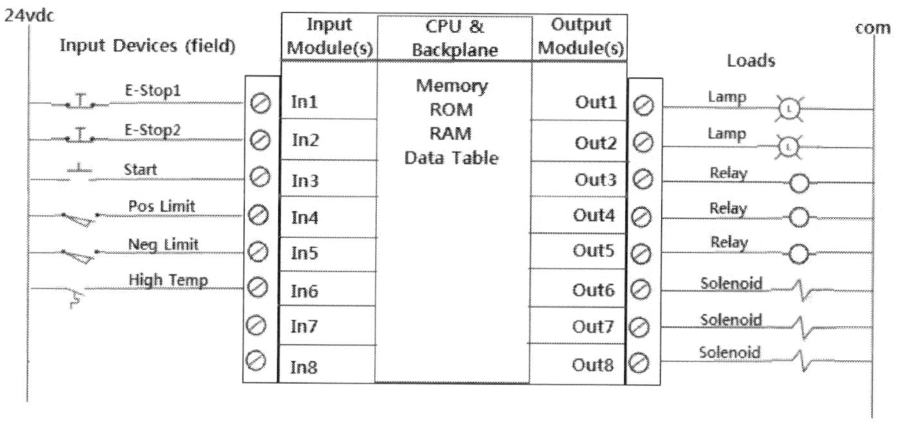

Digital modules process incoming signals as either "on" or "off"; these are commonly referred to as "***discrete***" modules. Other modules, both input and output, function to process or provide an "***analog***" type of signal in which voltage or current may vary over a range of values.

It should be noted here that your RSLogix 500 and RSLinx software are also used to program the Allen Bradley family of Micrologix PLCs, many of which are not modular but have ***fixed*** I/O with a specific number of input terminals and outputs. These are great to use when the need for input and output devices is more limited. However, many of the small PLCs having "fixed" I/O can also be expanded by the addition of another section of I/O modules.

Expandability is a key issue in today's PLC market. Indeed, it is now commonplace to link different racks each with its own processor or even several processors together in one rack and branch out to remote I/O with a single communication cable, as in the case with a CompactLogix controller branching to FlexLogix I/O blocks. For purposes of this series I will focus on the modular units that are common to the SLC 500.

At this point we've briefly looked into the architecture and hardware of a PLC system, in the next few chapters we'll get into the specifics of I/O module function, types of modules available, proper addressing and other important topics including instruction programming.

Chapter 2: Input & Output Modules!

Many different types of input and output modules are available that accommodate the great variety of input and output requirements on machinery and process equipment. There are 4, 8 and 16 point "discrete" modules that handle "on and off" types of functions, and also modules that will accept or produce "analog" signals. Examples of these types of input and output boards would be the 1746-NIO4V combination module which can accept either (+10 to -10 Vdc) or (+20 to -20 milliamps) as an input value and produces a (+10 to -10 vdc) output signal. Other examples of specialty modules are the 1746-HSCE, which accepts "encoder" counts to be used in the program for speed reference or positioning, and the 1746-NT8 module that accepts "thermocouple" inputs. Again, there is a wide variety of modular I/O from which to choose, all depending upon your specific needs.

The publication, ***CIG-WD001A-EN-P***, produced by Rockwell Automation, is a great reference source for I/O modules of every type and description for Allen Bradley PLCs. This can be downloaded directly from the Rockwell Automation website.

Regardless of the types of input and output modules that are utilized with a PLC system there are *five important functions* they all hold in common. If you keep these elements in mind it will help you to better understand the importance of what I/O modules actually accomplish.

- ***Termination*** points: I/O modules provide a convenient connection points for incoming (field devices) and outgoing (load) wiring.

- ***Threshold Detection***: Input modules provide input filtering to prevent unwanted "noise" and transient voltages from producing false data or signals.

Output modules pass the voltage or current signals necessary for their specific load and are free of undesired noise and transient voltages as well.

- ***Optical Isolation***: I/O modules provide effective isolation and shielding between diverse input/output voltages and the processor / backplane which operates at much lower, usually about (5 vdc) voltages.

- ***Translation***: Input modules process incoming signals – voltages and current, and produce a lower voltage signal, which the CPU processor will then interpret as "1s" or "0s" for use in executing the ladder logic program. Output modules receive data signals from the CPU, again "1s" or "0s", and pass or produce a signal, voltage or current, which is needed to drive a specific output load such as a solenoid, relay or analog device such as a I/P module or proportional valve.

- ***Indication***: LED indication for the terminal point which gives a visual on whether the input is on or off.

Here once again, is our block diagram showing these *essential functions*:

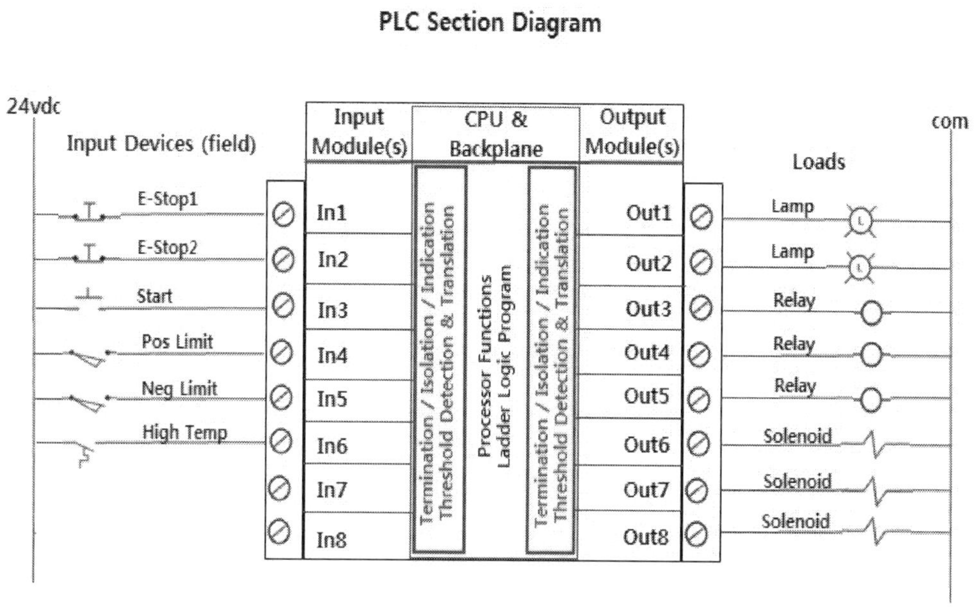

PLC Section Diagram

24vdc
Input Devices (field)
E-Stop1
E-Stop2
Start
Pos Limit
Neg Limit
High Temp
Input Module(s)
In1
In2
In3
In4
In5
In6
In7
In8
CPU & Backplane
Termination / Isolation / Indication Threshold Detection & Translation
Processor Functions Ladder Logic Program
Termination / Isolation / Indication Threshold Detection & Translation
Output Module(s)
Out1
Out2
Out3
Out4
Out5
Out6
Out7
Out8
Loads
Lamp
Lamp
Relay
Relay
Relay
Solenoid
Solenoid
Solenoid
com

The main "*translation*" function of an input module is to convert incoming voltages from "sensory" devices or switches into the "1s" and "0s" which are used in digital processing. Most PLC manufacturers produce input modules that are designed for 120 VAC, 240 VAC, 24 VDC, 5 VDC and other input signals as well. Output modules are designed to take the "1s" and "0s" from the processor and produce the appropriate output voltage or current – depending on how the board is configured.

The next two diagrams give a *simplified* depiction of how these functions are accomplished with a single input point and a single output point. Note that these modules also have voltage regulation and filtering, provided by zener / capacitor circuitry that deliver the necessary "*threshold detection*" for the incoming voltage. Most input and output modules are rated for a range of incoming voltages, such as the 1746-IA4, -IA8 or –IA16, whose operational voltages are rated at 85 to 132 VAC. Any incoming voltage within this range will be considered a usable signal and be processed by the input module, which in turn provides a usable signal for the CPU.

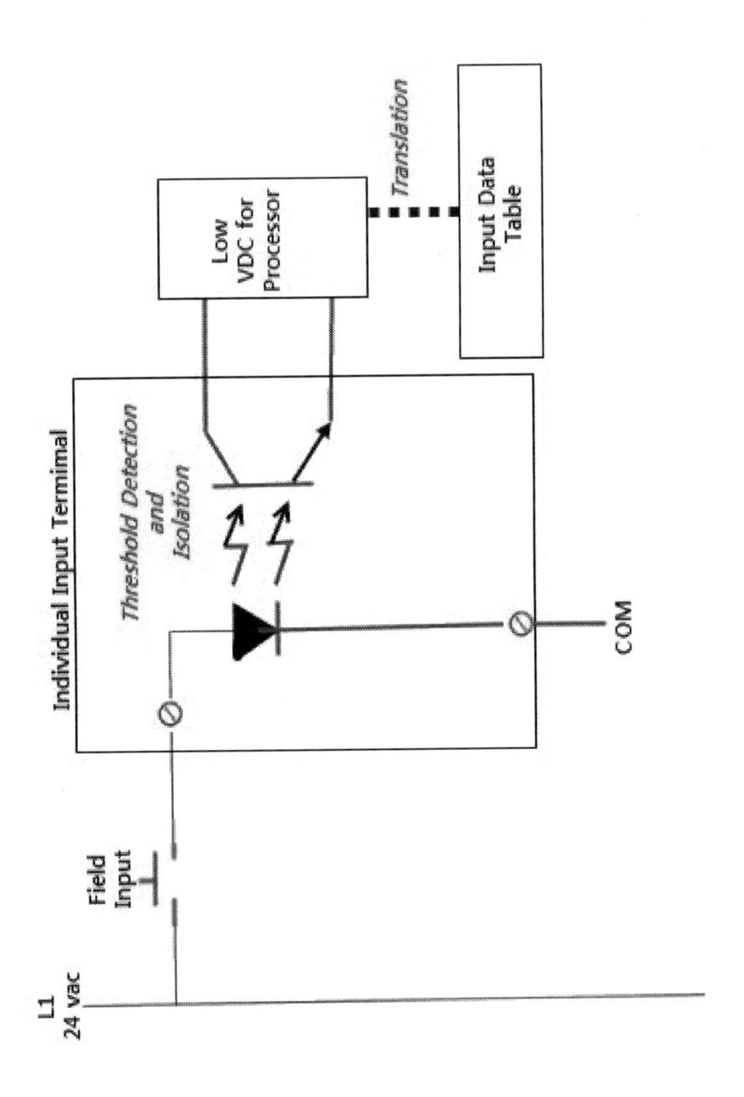

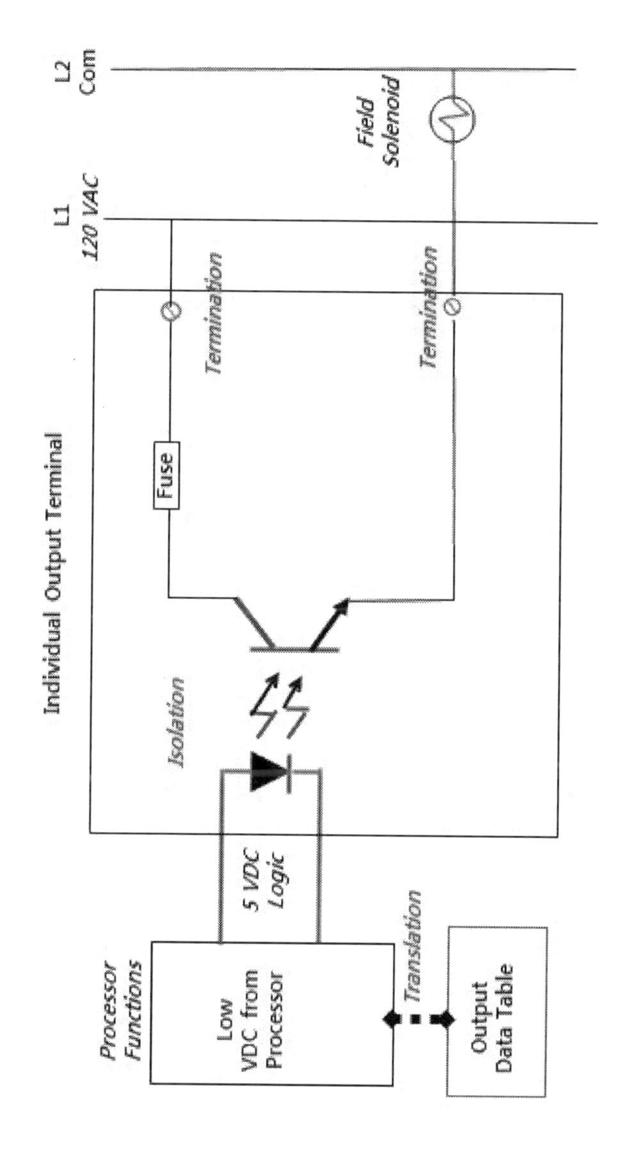

Individual Output Terminal
L2
Com
L1
120 VAC
Field
Solenoid
Termination
Termination
Fuse
Isolation
5 VDC
Logic
Processor
Functions
Low
VDC from
Processor
Translation
Output
Data Table

There are several types of discrete I/O modules available that include 4, 8, 16 or even 32 terminal points. There are also modules that interface with AC, DC, and TTL voltage levels. Output modules are available with solid-state and relay outputs, both in AC or DC voltage ratings.

Field Wiring for PLC & Devices:

In concluding this chapter I wanted to show some basic diagrams for the control wiring and protective circuits for both a PLC rack (Power Supply, CPU & backplane) and the I/O sections of a PLC system. This is representative of "typical" control wiring which *can and will vary,* according to your specific equipment.

I/O modules for DC voltage are categorized as "***sinking***" or "***sourcing***". For a "***sinking***" input or output module - the negative potential (0 vdc for example), is provided *through* the module circuitry. With "***sourcing***" input & output modules, the positive potential (+vdc) is provided or "sourced" *through* the module circuitry.

Sinking I/O Modules: Provide a path for conventional current flow to Common (-)vdc
Sourcing I/O Modules: Provide a path for conventional current flow to the "field device"- either a input device or load.

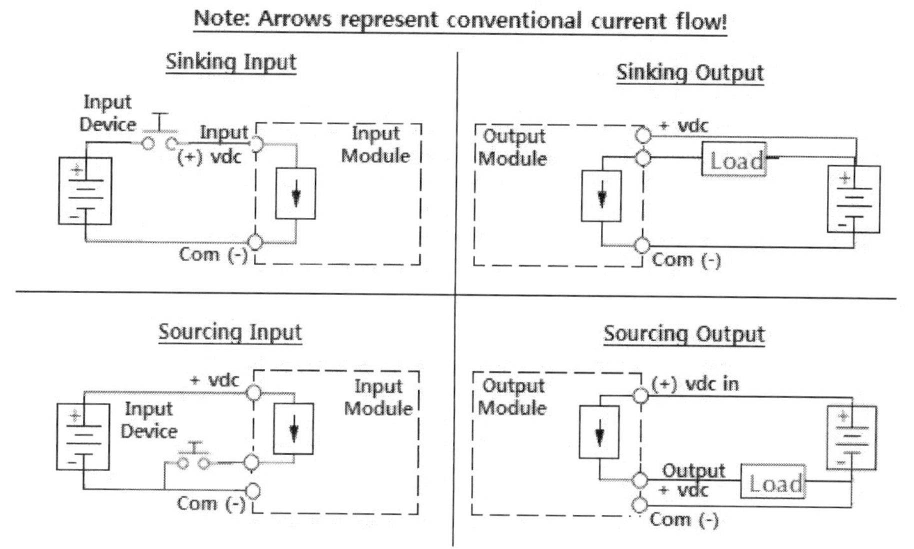

Note: The "field device", whether a discrete input or a load, is "sourcing" for a "sinking module" and "sinking" for a "sourcing module"!

The following diagram illustrates "typical" control wiring for providing power to PLC and the I/O modules using a "master" control relay.

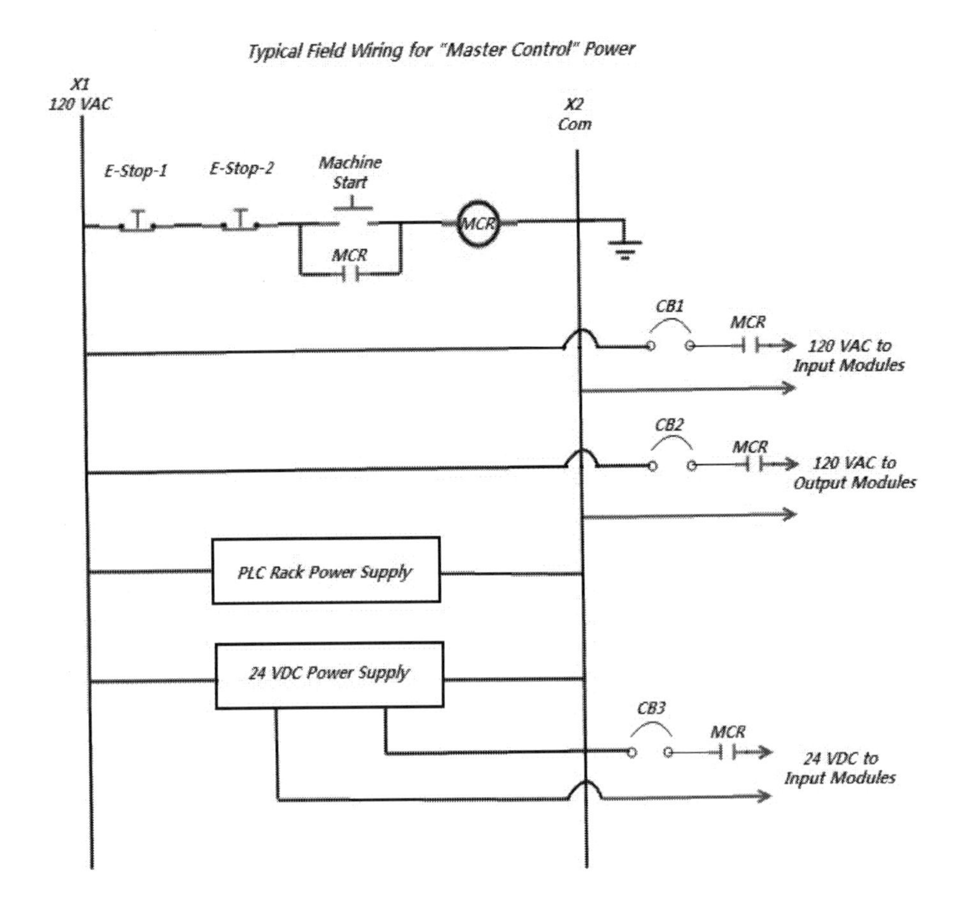

Digital Input Termination - 8 point

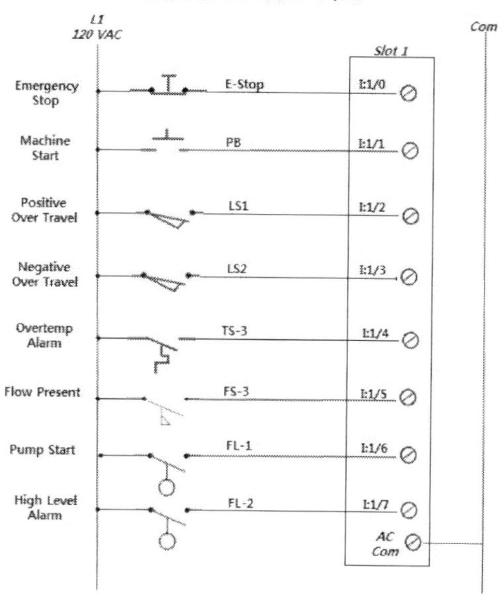

Digital Output Module - 8 point

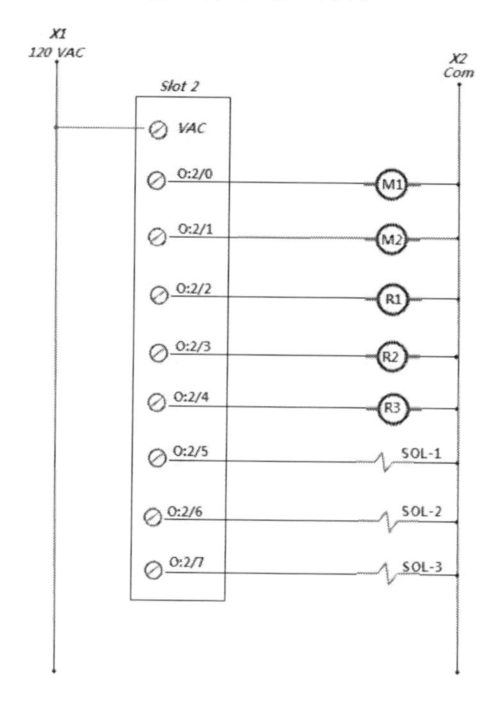

Chapter 3: Addressing, Memory and Scan Routine!

As you have seen from the previous chapter, you enjoy great flexibility in putting together a PLC system. This is a process of determining what tasks the equipment must perform, what input or "sensory" conditions will control and monitor the process when "running", what the equipment must do if these conditions are not met. At this point you are able to choose which PLC to build and will select the appropriate power supply, CPU, and the I/O modules- along with a backplane rack having the number of necessary slots. When this is accomplished you are ready to program the "ladder logic" that will cause the PLC to perform its task.

The objectives for this section are to:

- Understand module location-dependent addressing.
- Understand user-definable addressing.
- Understand how input and output data is "stored" in processor memory.
- Understand the PLC's scanning and execution routine.

A major component of programming certainly *begins* at the level of "addressing", as the processor must know the exactly which slot and terminal point each signal or "condition" is entering the PLC and which outputs, wired to their respective field devices, to turn on or off. Other outputs may only be "internal" bits which turn on ("1") or off ("0"), still other outputs may move a number into a specific register to be compared or used in some way by your program. In any case, addressing must be exact and begins as you start configuring your project.

Configuration for a PLC Project:

Below is an example of the RSLogix 500 application software and the screen used to "select" the CPU, as well as the input and output modules you will use in your system. At this point, you are designating not only what those I/O modules are, but also the specific "slot" where they will reside.

This screen opens in the RSLogix software when you select "new" under the FILE menu.

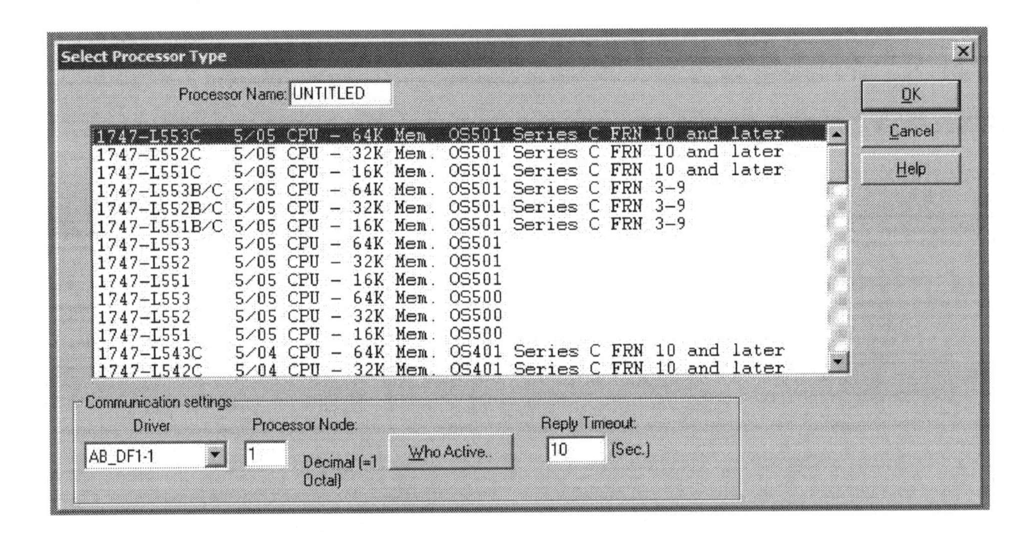

After selecting the CPU, select "I/O Configuration" and then the specific modules for the PLC as in the following example. "I/O Configuration" is in the project tree that resides along the left side.

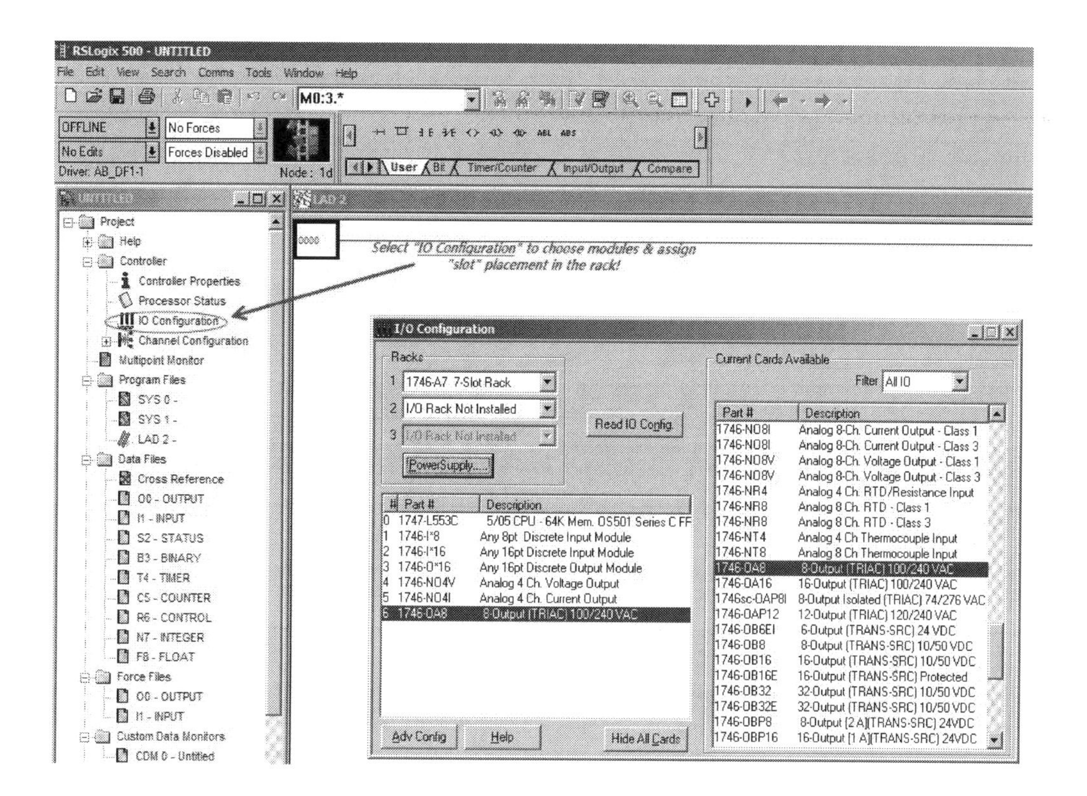

As you can see, I have selected a 7-slot rack with a SLC 5/05 model processor: a 1747-L553C, residing in slot 0. I have designated two input cards for slots 1 and 2, a 16 point output card in slot 3, two analog modules in slots 4 and 5, and a "triac" output card in slot 6. As you "double-click" from the selection menu on the right, the specific module will fill in your rack space on the left.

If you click on the "*Power Supply*" button a screen will appear that lets you know if you have exceeded the rating of your configured power supply and serves as an aid in selecting the one that will work with your project. Looking at the example below, if you "click" on the 1746-P1/P7, the message box will show that this rack, as configured, will overload the power supply. If you select the next one on the list, the 1746-P2 /P5 /P6, it indicates you will have a margin for the selected PS.

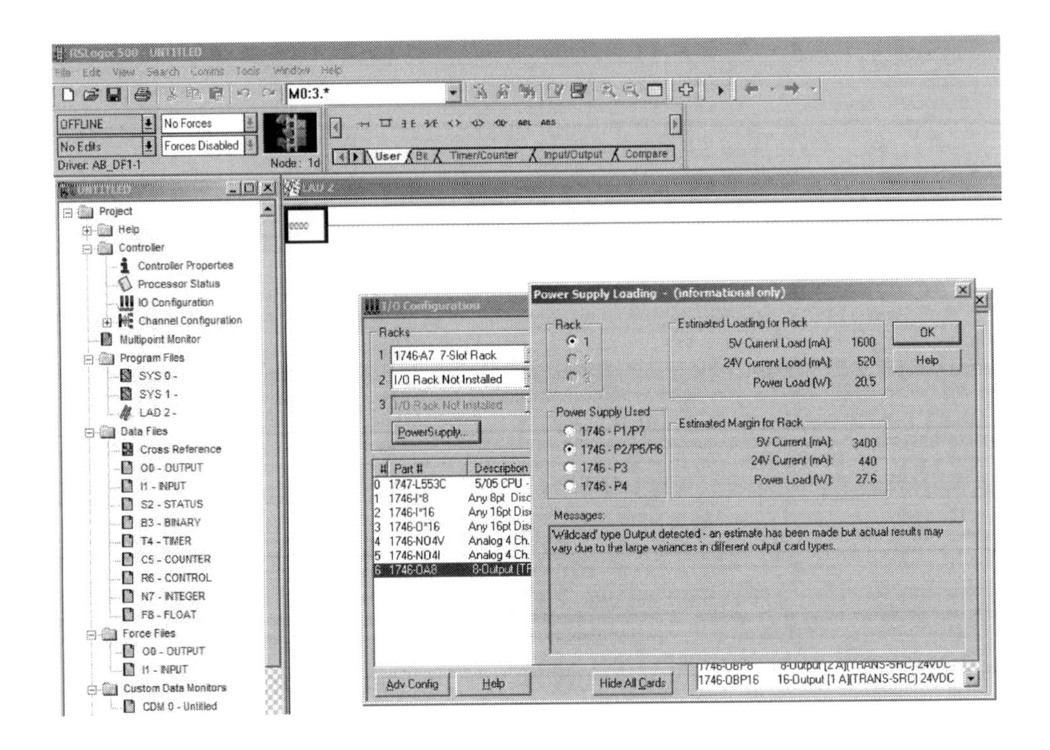

Addressing:

The configuration phase of a project is just the beginning of understanding the nomenclature of addressing. As you begin programming, each input & output terminal point will have a specific address, as will any internal bits, bytes, or words you use within your program. With that in mind, let's look at some of the basic concepts of addressing for the SLC 500 family of PLCs.

The addressing syntax and protocol used is referred to as, "*Module Location-Dependent Addressing*". It is basically the method most PLC manufacturers – including Allen Bradley, use to let the processor know "exactly" where to ***look for*** and ***send out*** the signals you have programmed into the ladder logic. Much like addressing a letter – where a zip code alone is not sufficient to send a letter to a friend, more detail is necessary. We need to be quite specific with the syntax used in addressing the inputs and outputs associated with your PLC project. Consider the following diagram for examples into the constructs of "*module-location-dependent* addressing.

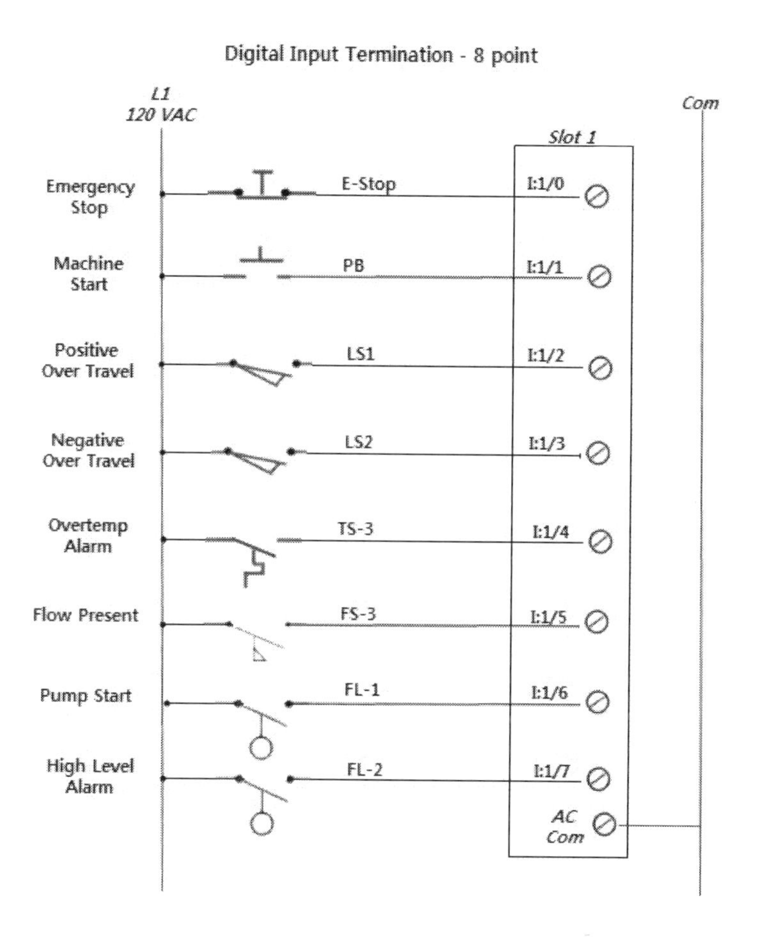

Addressing involves three parts as follows using the E-Stop input for an example:

- Data Type, "I" denotes actual physical "inputs" wired into the module.
- Module Location in the PLC rack. The "slot" number – in this example slot 1.
- Specific Terminal number on the module. In this example terminal "0".

As you enter the input conditions into your program that will become active when the E-stop button is pushed, we need to assign this address, "I:1/0", to every place in our program where the E-stop condition is used.

This same syntax is used for "outputs" Here is an example for our earlier diagram of an output card.

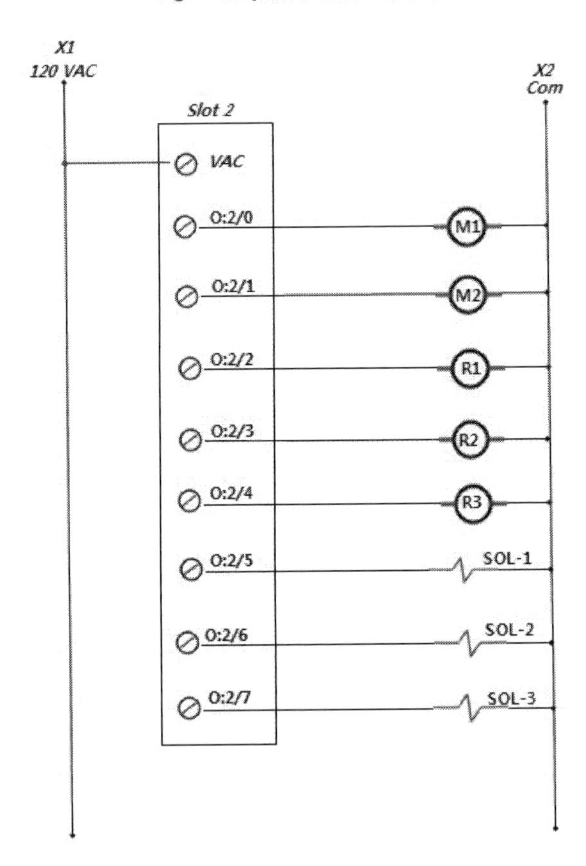

If we are writing a ladder program to turn on solenoid valves, SOL-1 and SOL-2, they would have to be addressed to "O: 2/5" and "O: 2/6" - wherever they occur in the program. Here is a short program in which the "start" pushbutton, "I: 1/1" would energize these solenoid outputs.

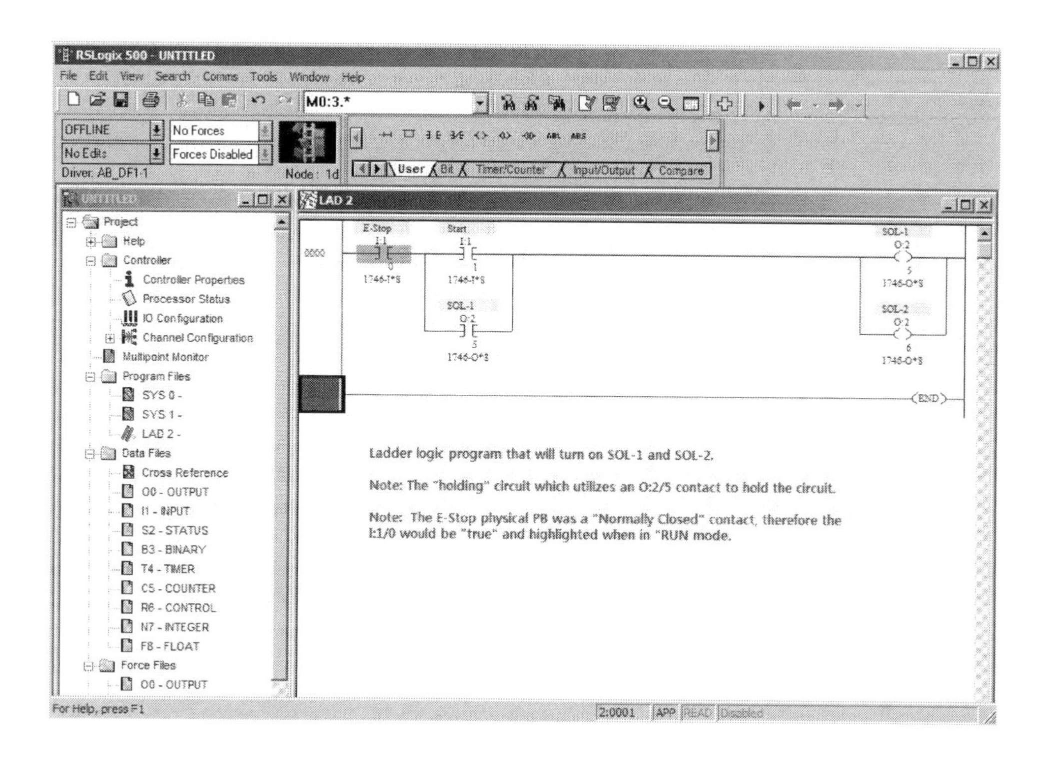

The Allen Bradley software highlights the "E-Stop" instruction, called an "**XIC**" or "***examine-if-closed***" instruction, whenever the program is loaded, in run mode, and the e-stop input on I:1/0 is *not pushed (note that it is a "normally closed" PB).* This condition causes the PB to pass voltage to the input module terminal, which causes the CPU to assign a value of "1" to a specific bit within a *data memory table*. So as the processor scans this input it *examines to see if it is closed*, if there is a voltage present – usually 3 to 5 vdc to the processor, it says the XIC instruction is "*true*" (closed) and assigns "1".

Here is the same ladder program after the "Start" has been pressed. Notice that other instructions and outputs are *highlighted* showing they are now "true", which means that the solenoids - in some remote physical location on the equipment, are energized.

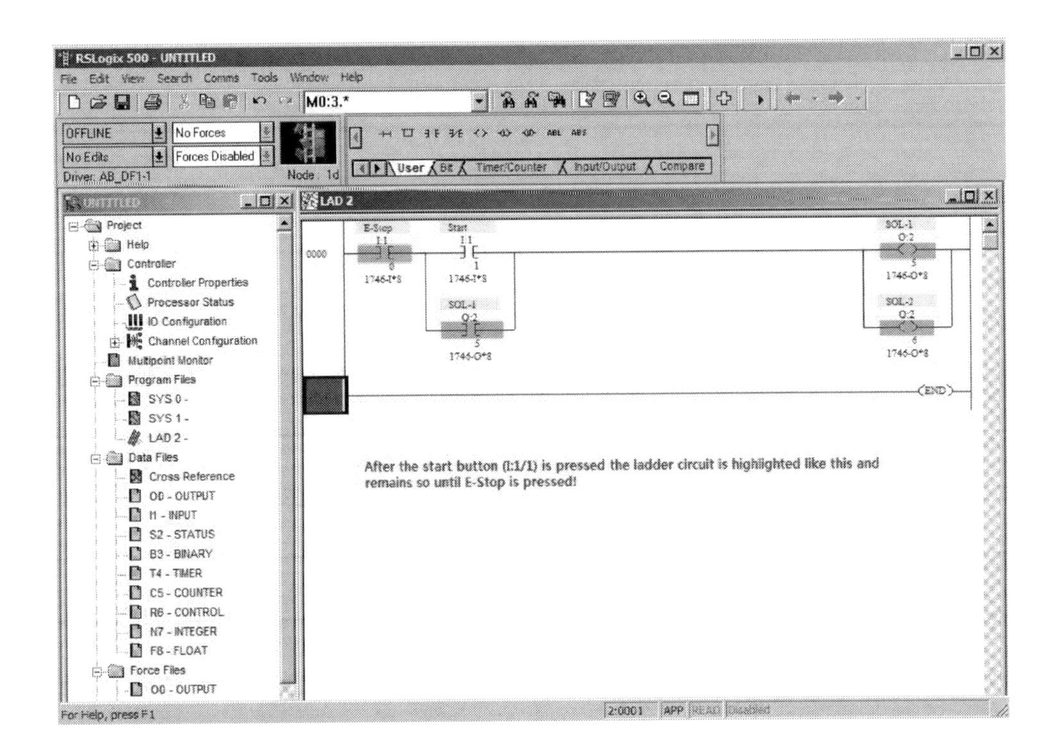

It's important to remember that the "contacts" shown on ladder logic programs – the XIC's in this example, <u>*are not physical contacts*</u> at all and exist only in the program itself. An XIC such as I:1/1 can be used many times throughout a program, although this is usually not the most efficient way of programming – but more on good programming practices are considered in later chapters.

If you look closely at the above example, you will see a "<u>*project tree*</u>" on the left side of the application window. Under the folder heading "<u>*Data Files*</u>", you can see the different data types listed beginning with the "output" or "O0" table and ending with a data type "F8", which is used for "floating-point" numeric values that can contain a "decimal" place. "N7" is a table that will hold "integer" values usually from 0 to 32767, the largest signed number a 16 bit word can hold. The "N7" file, like the "F8" file is a *word-level* file. These file-table numbers (3-8) are the "default" files numbers for Allen Bradley, but additional file-tables (9-255) can be configured in a project. These additional files can be configured as the same types already listed (IE: B9, B30, or N12, N15, etc.) or configured for ASCII or String values.

Here is a chart showing *general* addressing syntax for Data Files:

Allen Bradley File Types (default)						
Data File Type	Data File Number	Delimiter	Slot or Word number	Delimiter	Bit or Terminal Number	Description
O	0	:		/	Terminal #	Physical Outputs from output terminals
I	1	:		/	Terminal #	Physical Inputs to input module terminals
B	3	:		/	Bit #	Binary bit level status,virtual conditions or relays
T	4	:		/	Control Bits: 13 (DN), 14 (TT), and 15 (EN)	Timers
C	5	:		/	Control Bits: 10 (UA), 11 (UN), 12 (OV),13 (DN), 14 (CD), and 15 (CU)	Counters
N	7	:		/	Element (16 bit word) or Bit#	Integer Values "1-word" element
F	8	:		/	Element (16 bit word) only!	Floating point data "2-word" elements

Now let's attempt to put all this together! Right now we'll look at input, output, the B3 or "*binary*" file addressing examples and then follow with integer and floating point file addressing. It is important to understand each file type because of minor differences in syntax.

Here is the chart for the first three file types, all of which address to "bit" or "terminal point" level:

Data File Type	File Number	Delimiter	Element(word) or Slot #	Delimiter	Terminal or Bit #	General Description
I	1	:	Slot #	/	Terminal #	Physical Inputs to input modules
O	0	:	Slot #	/	Terminal #	Physical Outputs to output modules
B	3	:	Element #	/	Bit #	Binary, virtual conditions or relays

The "I" input and the "O" output files have been discussed already. You have probably noticed that it isn't necessary to include the "***file number***" when you address *these* instructions in your program. This is because these data-file types are associated with an *actual physical module* assigned to a physical slot in the PLC rack. The other file-table types reside in *program memory* within the CPU and since we can configure additional tables of these file-types it is necessary to assign and use *file-numbers* so bits or words can be found in a "user-configured" file such as N12 or B30.

The syntax necessary for type "I" and "O" instructions is simply **"file type": "slot#"/ "terminal",** for example I:1/7 for slot 1 and terminal . I will show examples of this in the following chapter, but as you "drag" an instruction element into a ladder rung or click on an instruction that is not addressed – it will highlight the instruction with a "?" indicating the need for an address.

You can simply type in the address at this point. Sometimes the ladder program will show the "*formal*" or complete address in this format: I:1.0/7 which is showing the "*word delimiter*" (a period) and also the specific word "0" before the terminal number.

Here is an illustration of the steps when using this "basic" type of addressing!

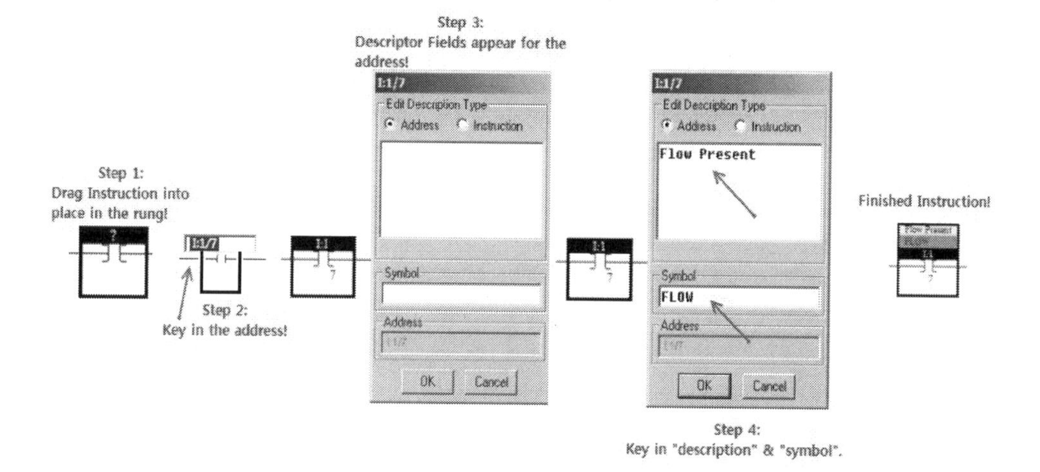

Addressing the B3, N7 & F8 Data Files:

The B3 file table is composed of 256 one word elements (0-255) – each word consisting of 16 bits. If using and addressing specific bits in B3 file there are two methods of addressing.

Method 1: **"file type/number": "element" / "bit #"** which designates both specific "word" and "bit #". Example: **B3:3/14** (designates bit 14 in word 3 in the B3 table)

Method 2: **"file type/number" / "bit #"** where the element or word is not designated but the bit is addressed by sequential number. In the "B3" table there are a total of 4096 bits (256 x 16 = 4096) any one of which (0 through 4095) can be used and addressed in this fashion. Example: **B3/75** or **B3/4095**

Here is an example of several file types and correct addressing:

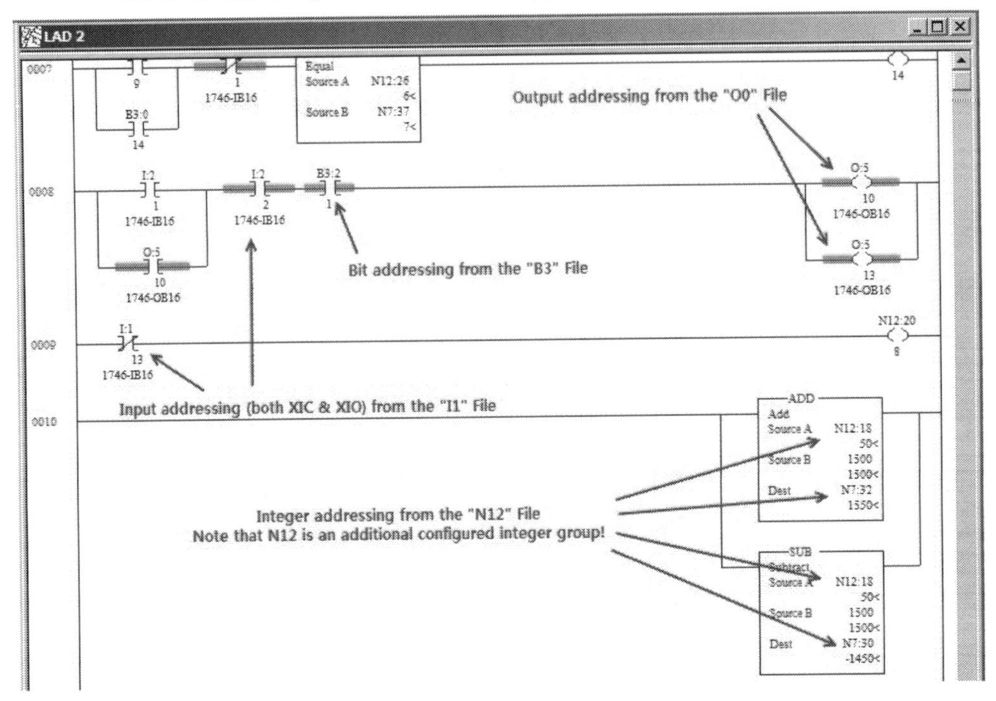

The "N7" integer file allocates one word (16 bits) to a number so will handle -32767 to +32678. Bit 15 is needed for the "sign", either positive or negative.

The "F8" or float file type allocates two words per address and so will accommodate much larger numbers.

For all of these "file-types" there is an easy method to view the exact value or the status of a particular input, output or bit. Simply "double-click" on the file you want to view from the project tree and it will bring out the word elements of that file. You can toggle between "binary" and "decimal" format and other formats as well.

Here we view the "B3" data file:

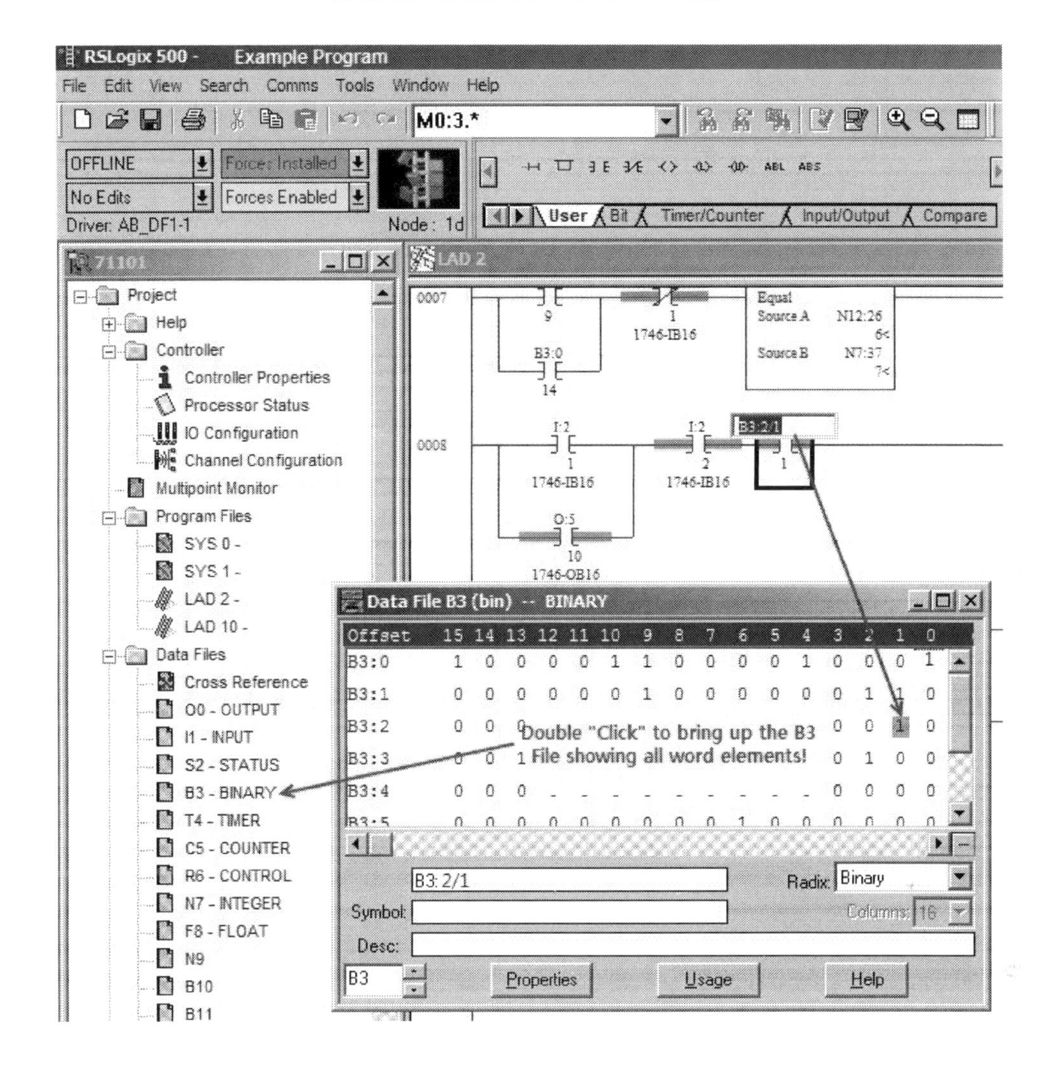

Here is a view of the integer "N7" data file in "decimal" format –

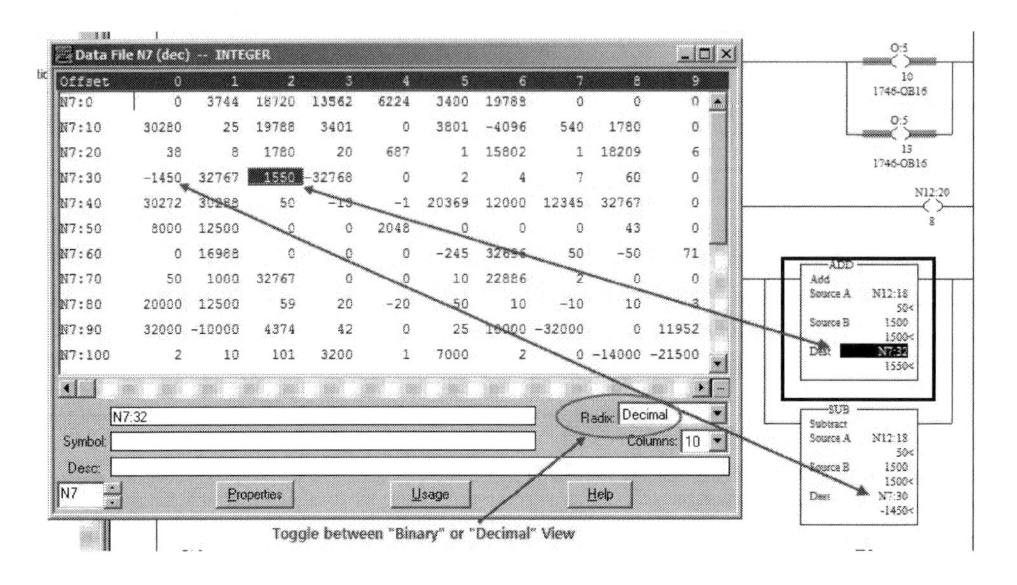

The above example show the "N7" file in the "Decimal" format so remember that each number you see, such as the N7:32 value of 1550 (highlighted), actually occupies a 16 bit word. Here is the same screen capture switched to the "binary" view. As you can see, a full word is used to represent the number 1550.

Binary view of the "N7" Data File!

Note that the N7:32 address references a 16 bit "word" element for the value of "1550".

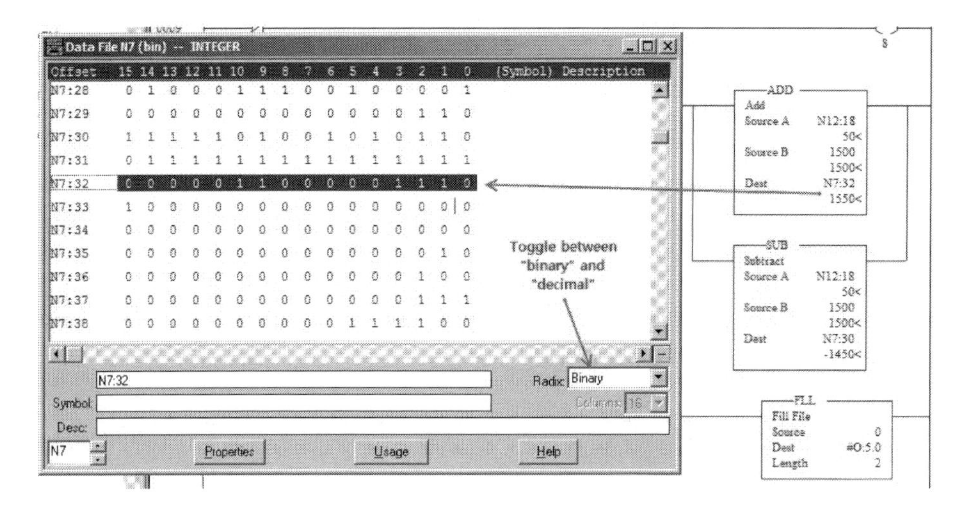

Remember that the "N7" file can be addressed and accessed at the ***word / element*** level or the ***bit*** level. Here are examples of both!

N7:3 would be interpreted as: "***N7 file, element 3***" and use whatever number is in the *16 bit word*.

N7:3/5 would be interpreted as: "***N7 file, element 3, bit #5***".

The "F8" file addressing is basically the same as the "N7", but with some exceptions. First of all, each element in "F8" (0-255 elements) is composed of 32 bits (2-words) and are *not addressable* at the bit level.

Addressing Timers and Counters:

These important programming instructions are composed of 3-words allocated to each timer and counter. These will be covered in more detail in the next chapter as we look into actual programming. Here is a brief synopsis of addressing protocols for both the "timer" and "counter" instructions.

Timer Control Bits & 3-Word Allocation																
15	14	13	12	11	10	9	8	7	6	5	4	3	2	1	0	
EN	TT	DN							*internal use-not addressable!*							*Status Word 0*
Preset Value (PRE)																*Preset Word 1*
Accumulator Value (ACC)																*Accumulator Word 2*

Addressable Bits	Addressable Words
EN, "Enable" (Bit 15)	PRE, "the Preset Value"
TT, "Timer Timing" (Bit 14)	ACC, "the Accumlated Value"
DN, "Done Timing" (Bit 13)	

Counter Control Bits & 3-Word Allocation																
15	14	13	12	11	10	9	8	7	6	5	4	3	2	1	0	
CU	CD	DN	OV	UN	UA				*internal use-not addressable!*							*Status Word 0*
Preset Value (PRE)																*Preset Word 1*
Accumulator Value (ACC)																*Accumulator Word 2*

Addressable Bits	Addressable Words
CU, "Count Up Enable" (Bit 15)	PRE, "the Preset Value"
CD, "Count Down Enable" (Bit 14)	ACC, "the Accumlated Value"
DN, "Done" (Bit 13)	
OV, "Overflow" (Bit 12)	
UN, "Underflow" (Bit 11)	
UA, "Update ACC" (Bit 10)	

With these instructions, addressing is done at a bit level and depends on which bit you desire to use. For instance, if you are using a timer instruction and want some event to happen exactly when the timer begins its count, you would use (and address) the timer's bit 15, the (EN) or "enable" bit. If you wanted an event to occur when the timer "*timed out*" you would use bit 13, the (DN) or "done" bit.

Here is a sample program showing an example of "Timer" addressing. As you can see in this example the addressing syntax is "***File Type & number: timer # / bit***" such as **T4:12/13** or **T4:12/DN**. These can be addressed either way and are referencing the same bit!

Here are examples of "Timer" and "Counter" instruction addressing in ladder logic programs. Like the other file types, specific bits and words can be viewed by "*clicking*" on the file in the project tree.

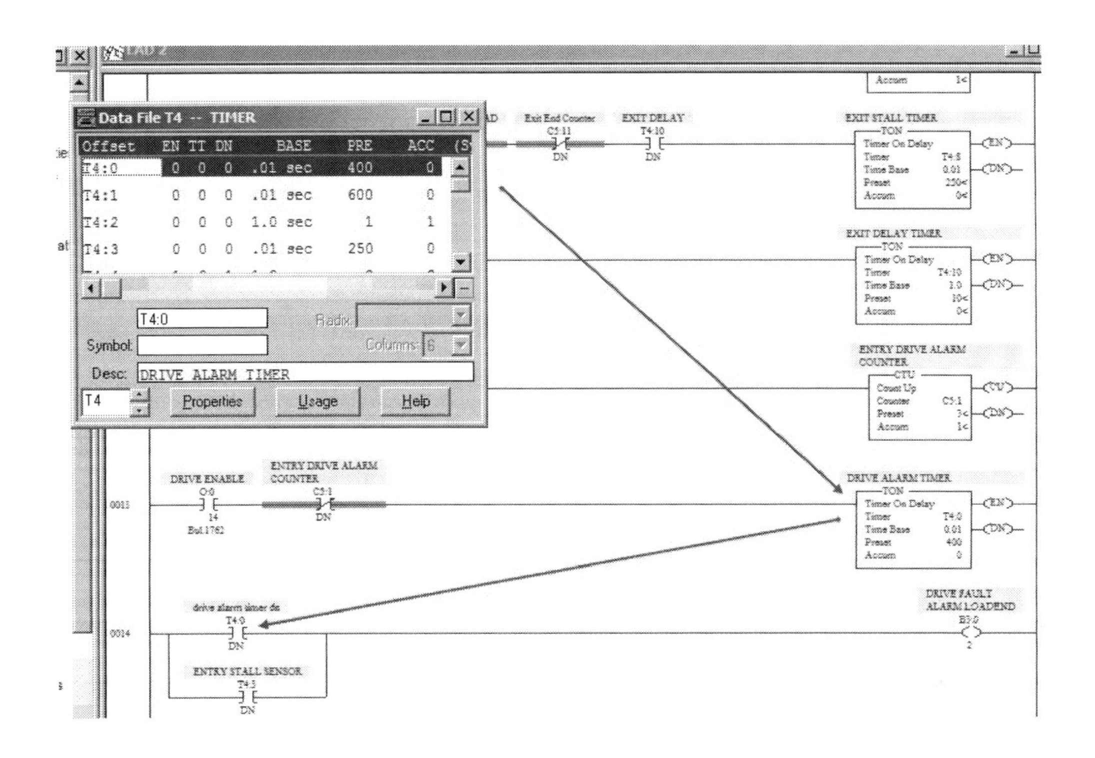
Data File T4 -- TIMER
Offset EN TT DN BASE PRE ACC
T4:0 0 0 0 .01 sec 400 0
T4:1 0 0 0 .01 sec 600 0
T4:2 0 0 0 1.0 sec 1 1
T4:3 0 0 0 .01 sec 250 0
T4:0
Radix:
Symbol:
Columns:
Desc: DRIVE ALARM TIMER
T4
Properties
Usage
Help
Exit End Counter
C5:11
DN
EXIT DELAY
T4:10
DN
EXIT STALL TIMER
TON
Timer On Delay
Timer T4:8
Time Base 0.01
Preset 250<
Accum 0<
EN
DN
EXIT DELAY TIMER
TON
Timer On Delay
Timer T4:10
Time Base 1.0
Preset 10<
Accum 0<
ENTRY DRIVE ALARM COUNTER
CTU
Count Up
Counter C5:1
Preset 3<
Accum 1<
CU
DRIVE ENABLE
O:0
14
Bul.1762
ENTRY DRIVE ALARM COUNTER
C5:1
DN
0013
DRIVE ALARM TIMER
TON
Timer On Delay
Timer T4:0
Time Base 0.01
Preset 400
Accum 0
drive alarm timer dn
T4:0
DN
0014
ENTRY STALL SENSOR
T4:3
DN
DRIVE FAULT ALARM LOADEND
B3:0
2

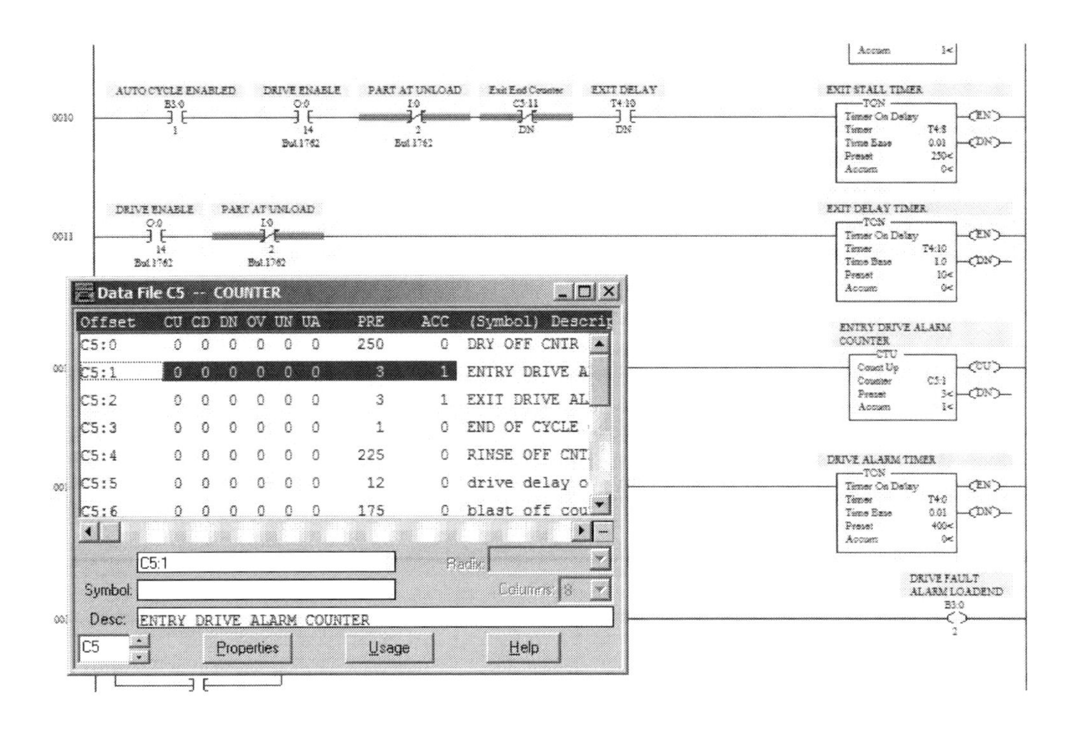
Accum 1<
0010
AUTO CYCLE ENABLED
B3:0
1
DRIVE ENABLE
O:0
14
Bul.1762
PART AT UNLOAD
I:0
2
Bul.1762
Exit End Counter
C5:11
DN
EXIT DELAY
T4:10
DN
EXIT STALL TIMER
TON
Timer On Delay
Timer T4:8
Time Base 0.01
Preset 250<
Accum 0<
EN
DN
0011
DRIVE ENABLE
O:0
14
Bul.1762
PART AT UNLOAD
I:0
2
Bul.1762
EXIT DELAY TIMER
TON
Timer On Delay
Timer T4:10
Time Base 1.0
Preset 10<
Accum 0<
Data File C5 -- COUNTER
Offset CU CD DN OV UN UA PRE ACC (Symbol) Descri
C5:0 0 0 0 0 0 0 250 0 DRY OFF CNTR
C5:1 0 0 0 0 0 0 3 1 ENTRY DRIVE A
C5:2 0 0 0 0 0 0 3 1 EXIT DRIVE AL
C5:3 0 0 0 0 0 0 1 0 END OF CYCLE
C5:4 0 0 0 0 0 0 225 0 RINSE OFF CNT
C5:5 0 0 0 0 0 0 12 0 drive delay c
C5:6 0 0 0 0 0 0 175 0 blast off cou
C5:1
Radix:
Symbol:
Columns:
Desc: ENTRY DRIVE ALARM COUNTER
C5
Properties
Usage
Help
ENTRY DRIVE ALARM COUNTER
CTU
Count Up
Counter C5:1
Preset 3<
Accum 1<
CU
DRIVE ALARM TIMER
TON
Timer On Delay
Timer T4:0
Time Base 0.01
Preset 400<
Accum 0<
DRIVE FAULT ALARM LOADEND
B3:0
2

We have only briefly shown examples of the "TON", an "on-delay" timer. Other types of timers, such as the TOF (off-delay) and RTO (retentive) and the use of the RES (reset) will be covered in subsequent chapters and in "***Advanced PLC Programming Concepts***", Book 2 of this series.

User-Defined Addressing:

This type of addressing is often referred to as "tag" based addressing. While not an option for the Allen Bradley SLC-500 family of processors, this type of addressing is common on newer style AB processors such as "ControlLogix" and "CompactLogix" processors.

With "user-defined" addressing, using the input module that follows, the I:1/0 could simply be referred to throughout the ladder logic program as "Stop" and the I:1/1 would simply be referred to as "Start". This is true of all the other inputs, outputs, bits and use of words within the ladder logic program. One thing to be careful with, should you be programming one of these "newer" controllers, is to not be random in the selection of tag names and to carefully document your "tags". One programmer could just as easily call I:1/1 by the tag "Purple" as he could "Start", which would of course become confusing. Normally confusion ensues when inputs or outputs carry very similar tags, such as "Start", "Start1", "Start1P", etc. Once again, not really an issue with programming an SLC 500, MicroLogix or PLC-5 controller since these will utilize the "*module-defined*" style of addressing.

Digital Input Termination - 8 point

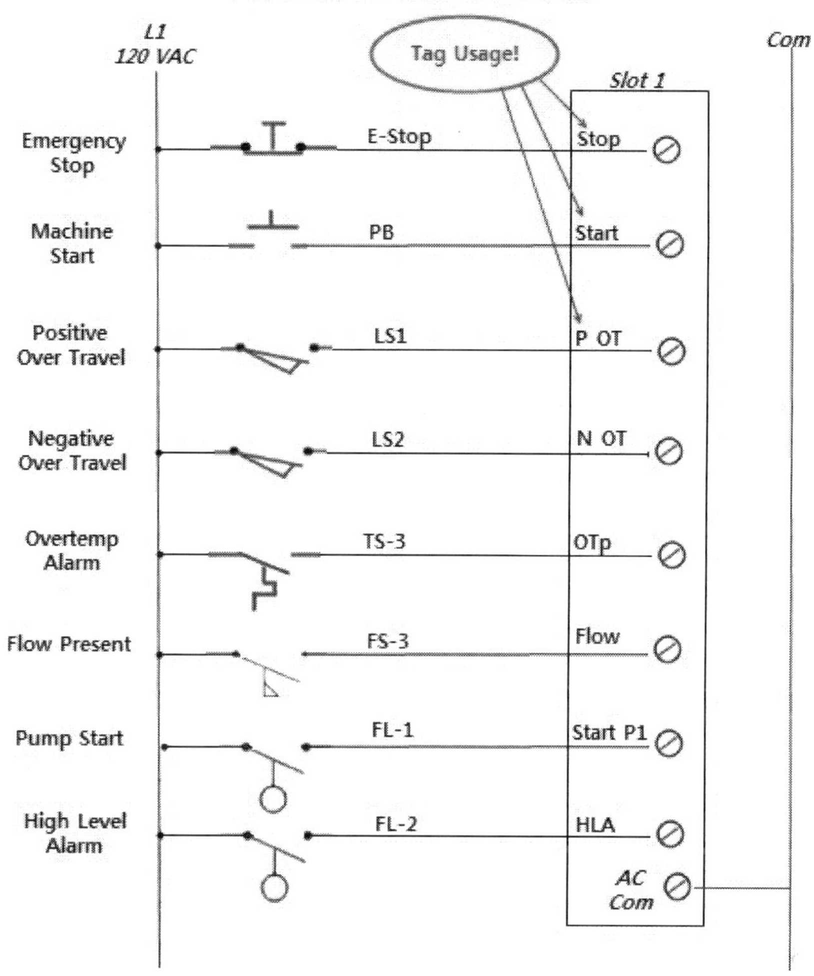

Scan Routine or Scan-Cycle:

The "Scan" routine on the PLC processor follows several definite and distinct steps every 5 to 50 milliseconds, depending on the size of the program and the number of I/O modules in use. They are as follows:

1. The Input Data Table is updated.
2. The program is executed.
3. The Output Table is updated.
4. Communication requests are processed.

Here is an illustration that summarizes a processor scan cycle!

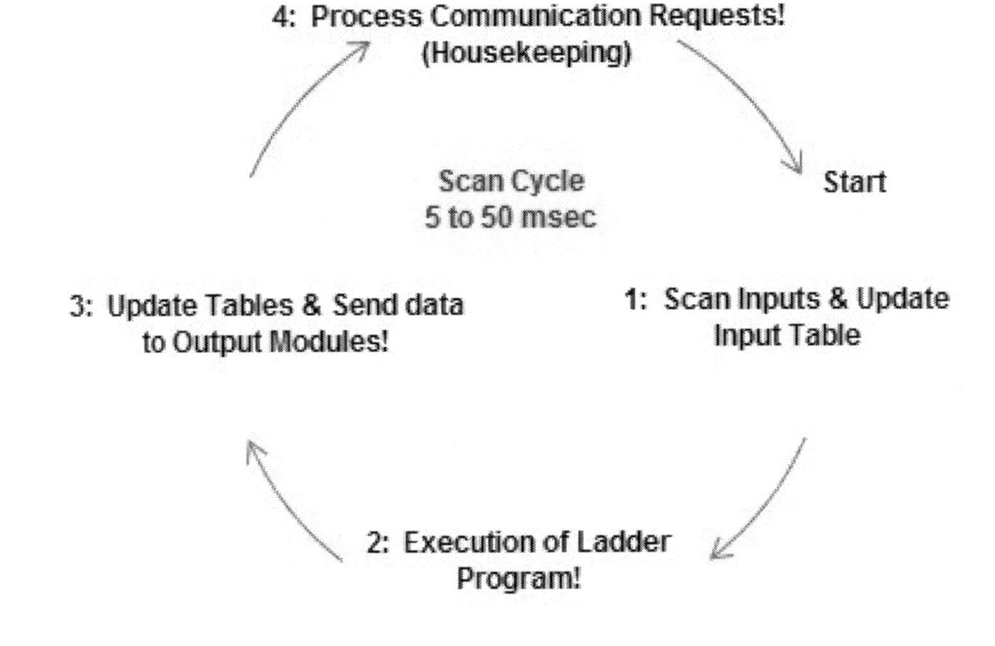

As the processor is switched to RUN mode, it begins this cycle. The first thing it does is to look at each input terminal to see if voltage is "present', and if so, to update the "input" table with a "1" or "0" in that specific bit location. If no voltage is present the bit is changed or is left at zero "0". The PLC scans all the inputs and updates the *input data table* in CPU memory. At this point the processor "executes" the ladder logic program based on these updated inputs and updates the data tables – both input and output. After the output data table is updated, the CPU sends these signals or information to the appropriate "output" modules. They will then either turn on or off, sending voltages, current or signals to specific field devices. During the final stage of the scan cycle the PLC processes external communication requests with devices such as HMI's (Human Machine Interface), touchscreens, and remote I/O.

Chapter 4: Programming Instructions!

So far we've looked into important details surrounding the types of hardware components that comprise a PLC, and how to address several different types of instructions. Of course the real objective here is to understand basic programming instructions that are used in writing a ladder logic program. We begin here with basic program instructions and include some discussion on how they are used.

The instructions for ladder logic programming can be divided into two broader categories: "***question – asking***" instructions and "***action – performing***" instructions.

"***Question – Asking***" instructions are those types of instructions that typically examine a condition to determine its state. Is it "on" or "off", "open" or "closed", or how does one value *compare* with another value.

These instructions would reside within this group.

–| |– *Examine-if-Closed* (ON): Think in terms of the "field device" or internal " bit, having this specific address, being checked to see if it is "closed" or "on".

–|/|– *Examine-if-Open* (OFF): Think in terms of the "field device" or internal " bit, having this specific address, being checked to see if it is "open" or "off".

–(OSR)– *One-Shot-Rising*: is a both a "*question-asking*" and a "*action-performing*" instruction. When conditions on the same rung that preceded the -OSR- become "true", it will set bit which is assigned to it (addressed) to "1" or "true" for one scan. On the next scan it goes back to a "0" status - even if the conditions ahead of it remain true.

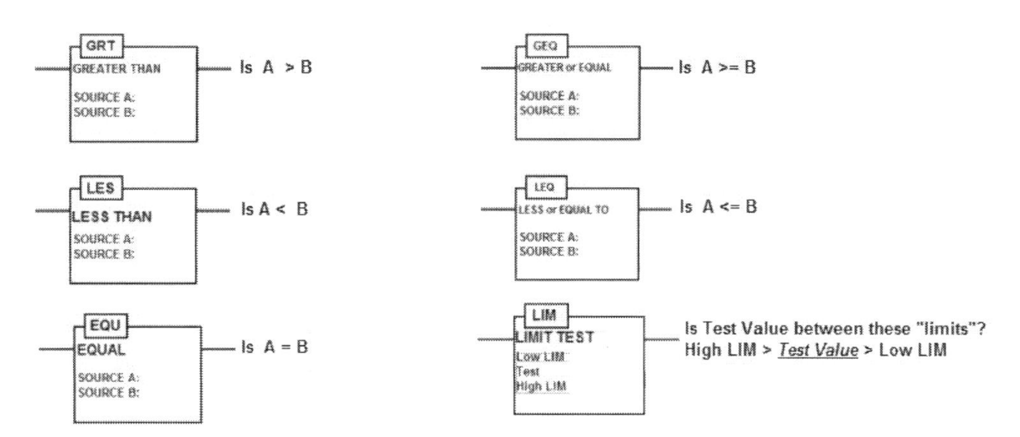

Note that the "comparison" type instructions are program elements that examine values, usually an N7 integer or F8 value, doing a comparison with another value. They may be a 16 bit register or a "constant" that you have placed in source A or B. If the condition is "true", then these instructions can provide logical continuity to the right side of the rung and activate an output or "action" instruction.

"***Action – Performing***" instructions are those types of instructions that typically perform a definite action. These can involve doing "math" calculations, moving or copying data into a different register, performing "timer" and "counter" functions, or simply turning an output terminal "on" or "off". The following instructions would reside within this group.

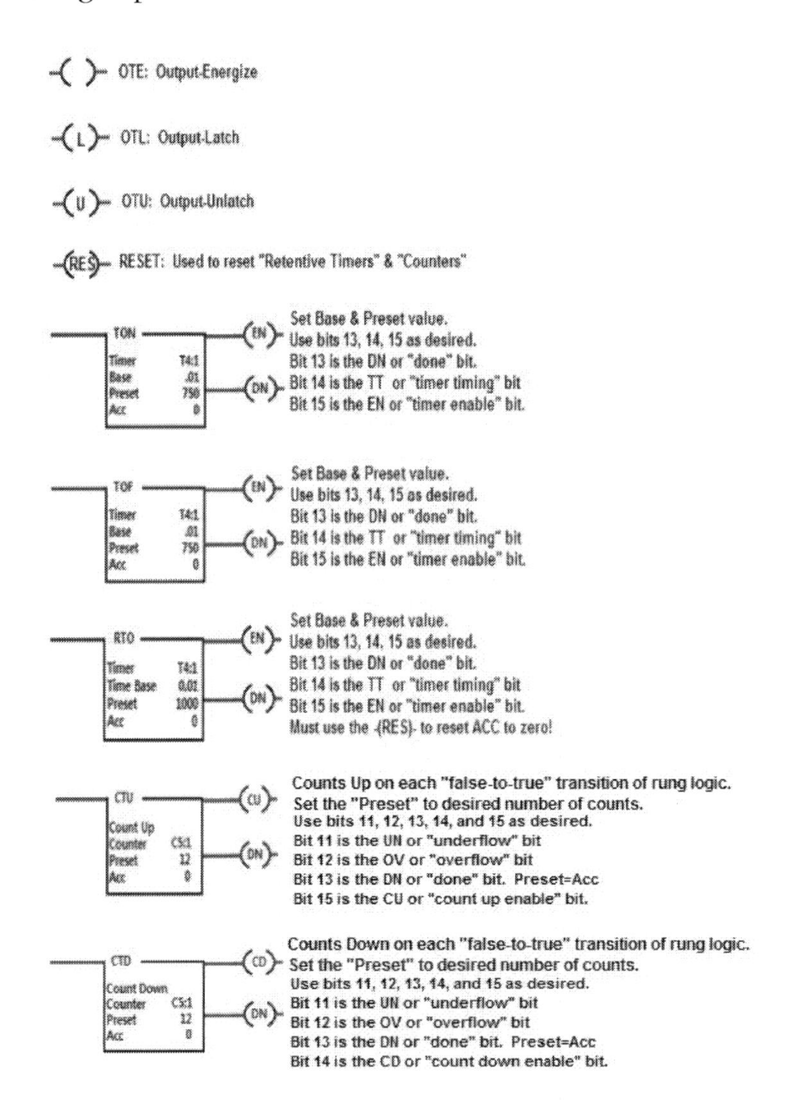

Register type "action" instructions are those that, when their preceding rung conditions are "true", will manipulate full registers (16 bits) of data by moving it into other locations, or that can perform math functions upon numbers in the registers you address.

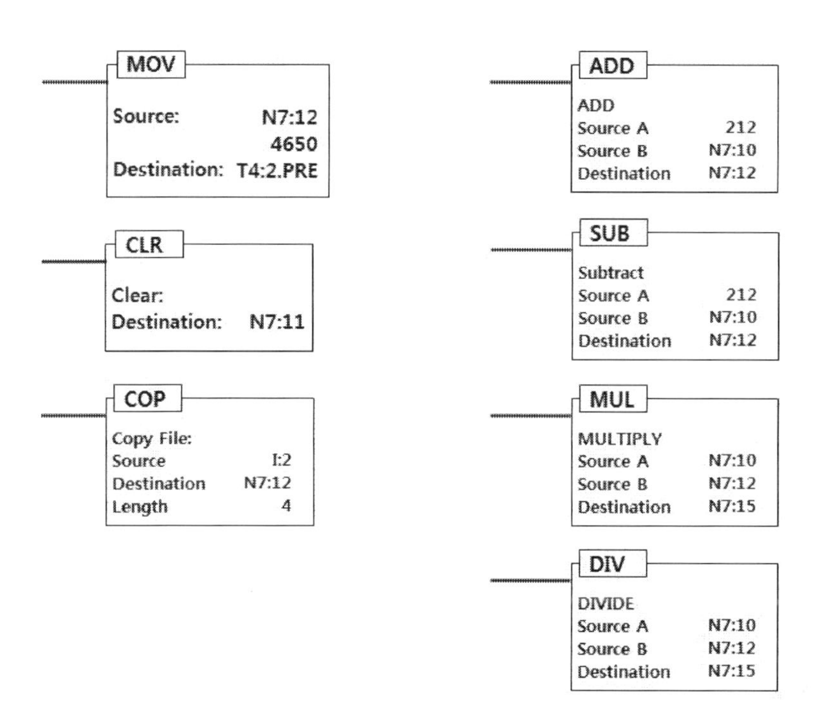

The "register" type of action instructions are important programming tools and find great usage in applications where analog values are used for input and output data such as motor speed, tension, torque or any number of other applications where conditions are constantly changing.

These instructions will be covered in greater detail in: "***Advanced PLC Programming Concepts***", Book 2 of this series.

We'll start into the specifics of using some of the basic programming elements, the XIC, XIO, and the OTE, and at the same time consider the importance of how these elements are "*arranged*" within a program rung. First a brief description of each:

Basic I/O Instructions

-| |- XIC: Examine-if-Closed (on)

-|/|- XIO: Examine-if-Opened (off)

-( )- OTE: Output-Energize

These instructions and many others can be "*pulled down*" into the program ladder structure; "*drag and drop*" would probably be more accurate terminology, from the top menu and placed into the desired rung. See the example below.

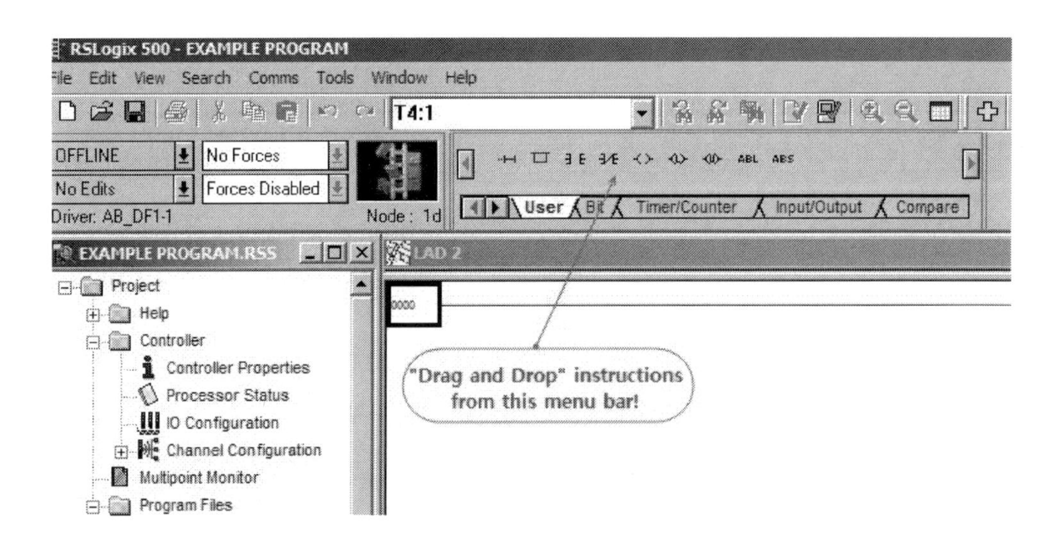

The "**XIC**" or "**examine-if-closed**" instruction becomes "true" if either the field device that is wired to it is "**closed**" or, if addressed to an internal bit in program memory, is that bit turned "**on**". Here is an example of both scenarios. Remember that the element being "examined" to see if it is closed is either the field device (limit switch, push button, pressure switch … etc.), or an internal bit address.

The "**XIO**" or "**examine-if-open**" instruction is "true" when either the field device that is wired to it is "**open**" or, when addressed to an internal bit in program memory that is turned "**off**". This instruction can sometimes be confusing but remember that, just like the "XIC", the "XIO" is examining a *field device* or an *internal bit* to ascertain whether it is *"open" or "off"*. A "true" condition is shown within the ladder logic screen as a "*high-lighted*" instruction. This indicates a "logical" continuity to whatever "*action-performing*" element is on the *right* side of the rung. When all the conditions that precede the OTE are "true", the OTE will be highlighted as well.

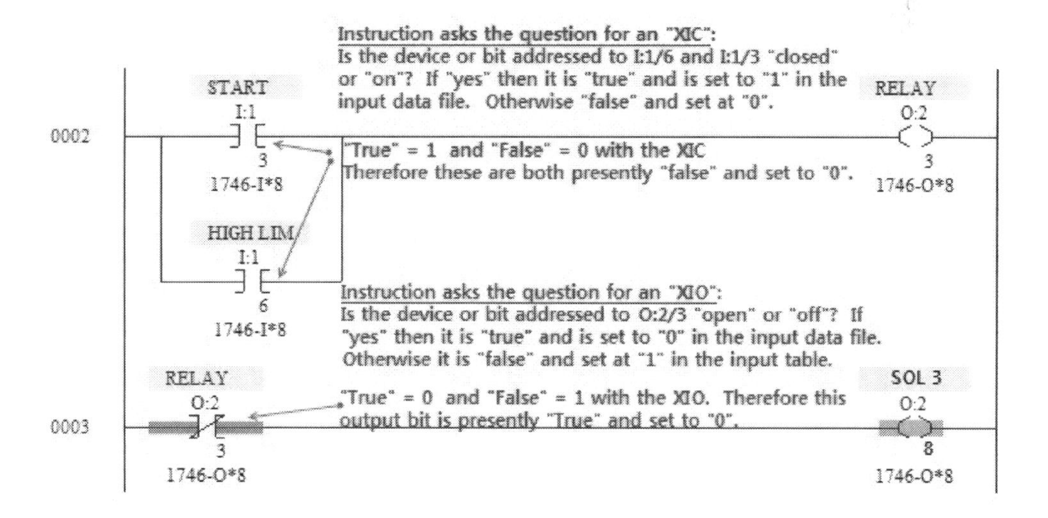

Whenever the Start pushbutton is pressed or if High Limit Temp is closed (see the mock-up PLC on following page) the I:1/3 or I:1/6 input will become "true" and pass *logical continuity* to O:2/3 which then will also become "true". This will cause the XIO addressed to O: 2/3 to change states and turn ***off*** the output to SOL 3 on rung 3. If wired to an actual relay, as in the example that follows, terminal 3, will output voltage to a control relay.

PLC Section Diagram

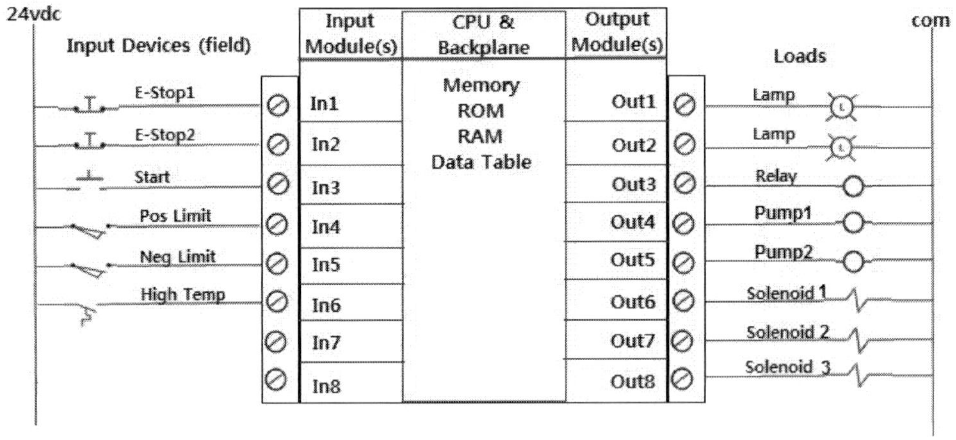

24vdc
Input Devices (field)
Input Module(s)
CPU & Backplane
Output Module(s)
com
Loads
E-Stop1
E-Stop2
Start
Pos Limit
Neg Limit
High Temp
In1
In2
In3
In4
In5
In6
In7
In8
Memory
ROM
RAM
Data Table
Out1
Out2
Out3
Out4
Out5
Out6
Out7
Out8
Lamp
Lamp
Relay
Pump1
Pump2
Solenoid 1
Solenoid 2
Solenoid 3

Chapter 5: Logical Operations & Arrangement of Instructions!

As you can see, instructions can be arranged in a great variety of ways depending on the conditions you desire for physical outputs to be turned on and off. If you are familiar with digital electronics, you have probably studied or seen the symbols for "*digital logic gates*", such as the "AND" or "OR" gate. Each of these varied conditions can be accomplished within our ladder logic program using these three instructions.

Using our "mock-up" PLC I/O, here are examples of each with the corresponding "truth" table.

PLC Section Diagram

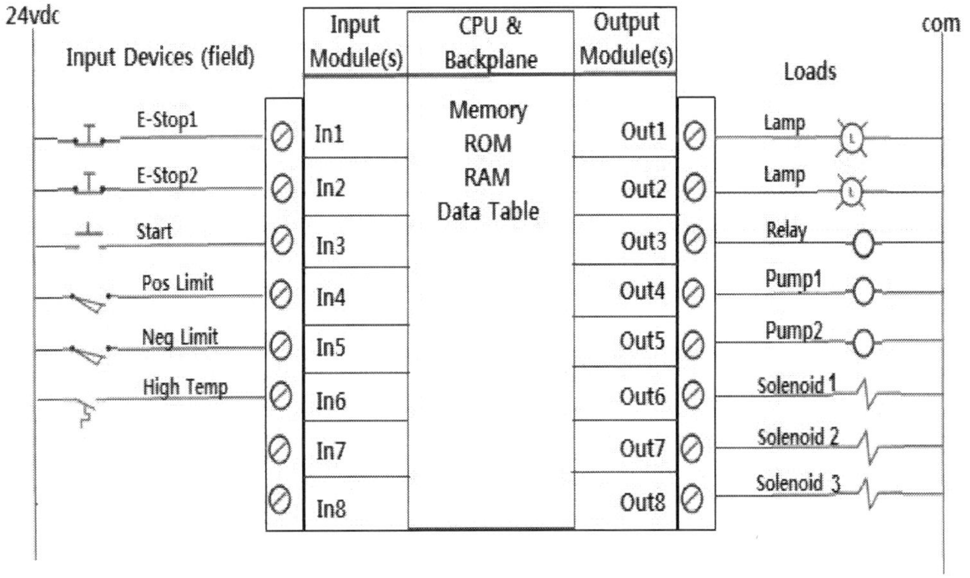

24vdc
Input Devices (field)
E-Stop1
E-Stop2
Start
Pos Limit
Neg Limit
High Temp
Input Module(s)
In1
In2
In3
In4
In5
In6
In7
In8
CPU & Backplane
Memory
ROM
RAM
Data Table
Output Module(s)
Out1
Out2
Out3
Out4
Out5
Out6
Out7
Out8
Loads
Lamp
Lamp
Relay
Pump1
Pump2
Solenoid 1
Solenoid 2
Solenoid 3
com

With "AND" logic, *all conditions* <u>must</u> be met or "true", before an action function is performed!

Logical "AND" instructions
If "HighLim" AND "PosLim" are ON - then turn on "Pump1"

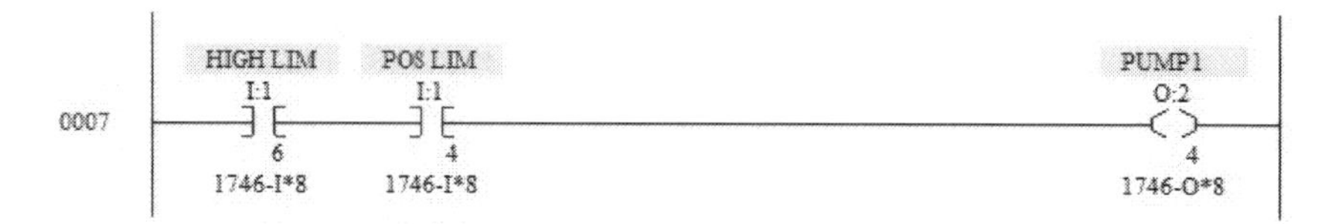

Truth Table for "AND" Gate				
Input	Input	Input	Output	
0	0		0	
1	0		0	
0	1		0	
1	1		1	

With "OR" logic, <u>either</u> condition (or both) must be met or "true" for the action function to be performed!

Logical "OR" Instruction
If "Start" OR "PosLim" is ON - then turn ON "Pump1"

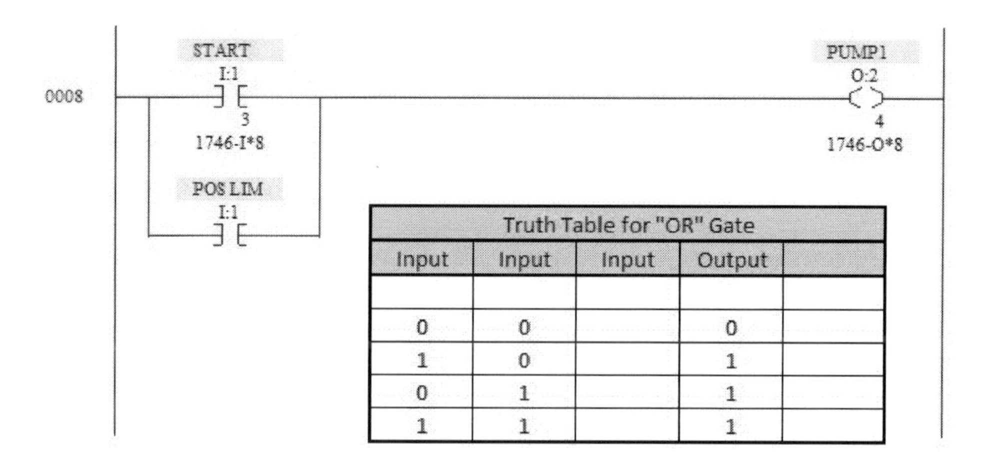

Truth Table for "OR" Gate				
Input	Input	Input	Output	
0	0		0	
1	0		1	
0	1		1	
1	1		1	

With the "NAND" and "NOR" rung arrangements, one output stops another from being able to become "true" – in a type of "safety interlock" fashion. Here are examples of both.

Logical "NAND" Instructions
If "Start" AND "HighLim" is ON then turn ON "Relay" and turn OFF "Sol 2"
(If "Relay" is ON, keep "Sol 2" OFF)

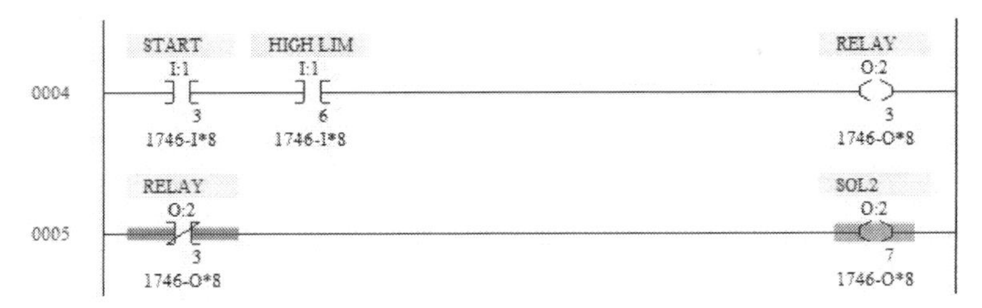

Truth Table for "NAND" Gate				
Input	Input	Input	"AND" Output	"NOT" Output
Start	HighLim	Relay	Relay	Sol 2
0	0	0	0	1
0	1	0	0	1
1	0	0	0	1
1	1	1	1	0

Logical "NOR" Instructions
If "Start" OR "HighLim" is ON then turn ON "Relay" NOT "Sol3"
(if "Relay" is on then NOT "Sol3")

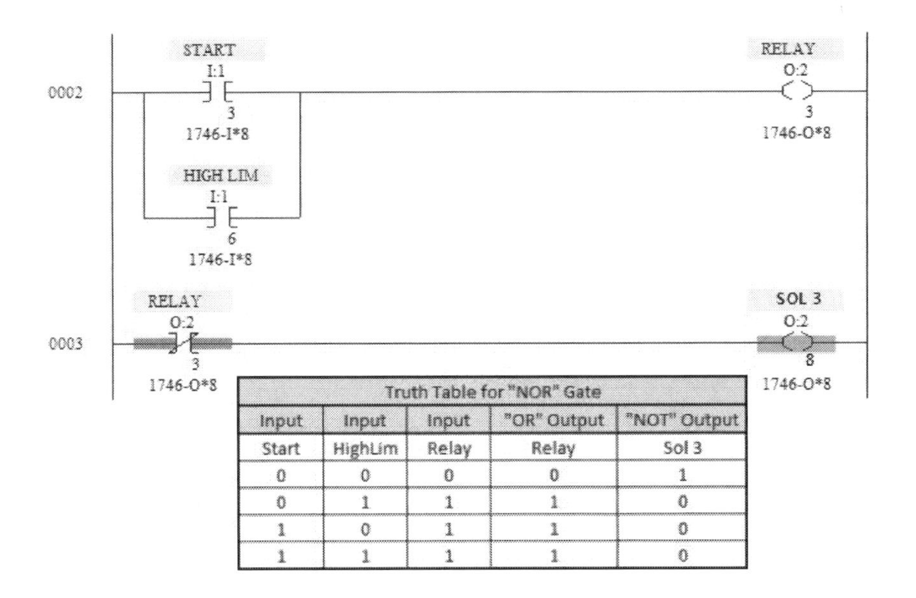

Truth Table for "NOR" Gate				
Input	Input	Input	"OR" Output	"NOT" Output
Start	HighLim	Relay	Relay	Sol 3
0	0	0	0	1
0	1	1	1	0
1	0	1	1	0
1	1	1	1	0

The final example of instruction element arrangement is the "XOR" or "*Exclusive Or*" arrangement. This allows for either of the input elements, but not both, to allow the action-performing OTE to become "true". While there may seem to be little need for this type of arrangement, when viewing only the small example below, you must consider the fact that these input conditions may be used throughout the program to control many additional outputs. It could be one of these outputs that the "XOR" is protecting.

Logical "XOR" Instructions
Allows either input (XIC -] [-), if true to turn on "Relay" output - but NOT both inputs!

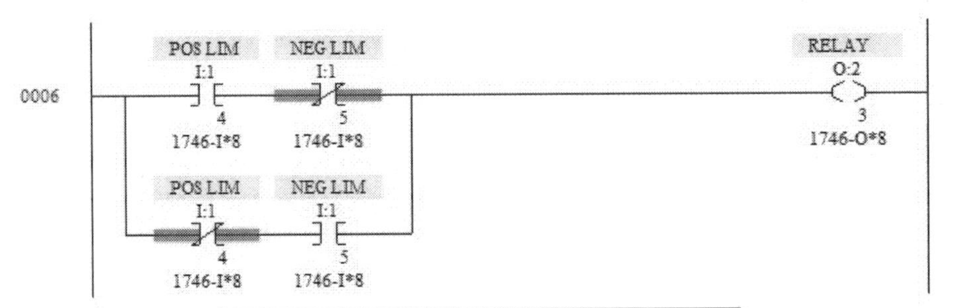

Truth Table for "XOR" Gate				
Input	Input	Input	Output	
0	0		0	
0	1		1	
1	0		1	
1	1		0	

Now is a good time to consider some of the other "basic" and often used "action-performing" instructions, the "latch output" –(L)-, "unlatch output" –(U)-, the "one-shot-rising" -(OSR)-instruction and basic timers, the "on-delay" (TON) and "off-delay" (TOF).

Here are the symbols and how they look in the actual ladder logic program!

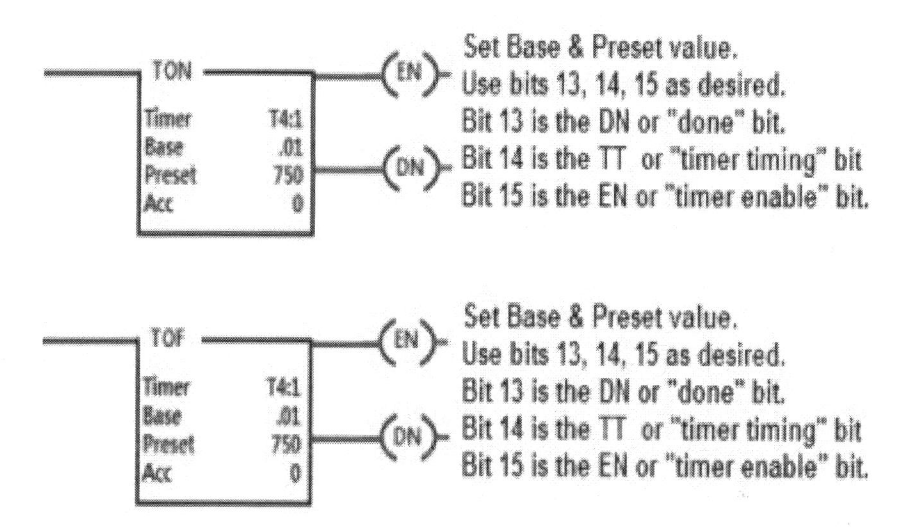

The ***OTL –(L)-*** output does just what is implied, when its preceding conditions are "true" it becomes true as well, and just like an OTE (output-energize), provides voltage to whatever output terminal and "internal bits", that share its particular address. The difference is that when preceding conditions change to a "false" status, the OTL remains on until it is unlatched by an OTU (unlatch) output having the same address.

The ***OTU – (U) -*** output, when its preceding conditions are met, will turn-on and unlatch the OTL *which shares its address.*

The ***OSR -]OSR[-*** is a conditional input type instruction which triggers an event to occur *once* by setting its bit location to "1" when its preceding conditions transition from "false" (0) to "true" (1). Once this scan is complete it sets the "bit" by which it is addressed back to "0" and remains so even if preceding conditions remain true. For this cycle to occur again, it is necessary for preceding conditions to once again transition from "*false to true*". This instruction is often used when a specific event needs to occur only one time during start-up.

Timers give us the available use of three different bits to use within our program and also the ability to set the amount of time delay we want to use. These descriptions are for the two basic types of timers, the **TON** and the **TOF**. An additional type of timer called the "**RTO**" or "*retentive timer*" will be discussed, along with "*counters*", in book 2, "*Advanced PLC Programming Concepts*".

The **TON** or "*on-delay*" timer uses **bit 13** or the "*done bit*" "**DN**", which is set to "1" when the timer *finishes* its time count. **Bit 14** is the "*timer-timing*" bit and is designated "**TT**" – it is set to "1" when the timer is *initiated and begins timing and goes to "0" when the DN bit is set.* The third bit we can use is the "*enable*", **bit 15**, which comes on when the timer initiates and begins timing – just like the "TT" bit. The difference is that the "**EN**" bit stays "on" until the rung conditions preceding the TON instruction transition back to "false". This resets the "EN" to "0" and the "accumulated" or ACC back to its start value.

Bit Description for a "TON"

Bit Designated	Is set ("1") when:	And remains set until one of the following conditions:
Timer Done Bit (DN) Bit 13	Accumulated value is equal to or greater that the Preset Value!	Rung Conditions go "false"!
Timer Timing Bit (TT) Bit 14	Rung conditions are "true" and the ACC value is less than the Preset!	Rung Conditions go "false" or the DN bit is set!
Timer Enable Bit (EN) Bit 15	Rung Conditions are "true"!	Rung Conditions go "false"!

The **TOF** or "*off-delay*" timer uses **bit 13** or the "*done bit*" "**DN**", which is set to "1" when rung conditions become "true". **Bit 14** is the "*timer-timing*" bit, designated "**TT**", and sets to "1" when rung conditions are "false". The timing is *initiated and "TT" remains set to "1" until rung conditions become "true" or the ACC value is equal or greater than the Preset value.* The third bit we can use is the "*enable*", **bit 15**, which comes on when rung conditions are "true" and stays on until rung conditions become "false".

Bit Description for a "TOF"

Bit Designated	Is set ("1") when:	And remains set until one of the following conditions:
Timer Done Bit (DN) Bit 13	Rung Conditions are "true"!	Rung Conditions go false and the accumulated value is greater than or equal to the Preset!
Timer Timing Bit (TT) Bit 14	Rung conditions are "false" and the ACC value is less than the Preset!	Rung Conditions go "true" or the DN bit is reset!
Timer Enable Bit (EN) Bit 15	Rung Conditions are "true"!	Rung Conditions go "false"!

Here is an example of how several of these instructions *might* be used within a program to control the following conditions and using our PLC diagram for inputs and outputs.

Example Program:

Conditions: After process start-up and upon reaching a specific temperature, turn on cooling system and maintain temperature within a predetermined range.

1. When Pump1 run cycle is started, ensure that all solenoid valves used for cooling are "off".

2. Open cooling solenoids (turn them on) when specific temperature is reached – determined be the "High Temp" switch.

3. Keep the cooling "on" for 15 minutes after temperature has fallen below the "high temp" setting. Then drop out cooling solenoids to allow reheating.

PLC Section Diagram

24vdc

Input Devices (field)	Input Module(s)	CPU & Backplane	Output Module(s)	Loads
E-Stop1	In 0	Memory	Out 0	Lamp
E-Stop2	In 1	ROM	Out 1	Lamp
Start	In 2	RAM	Out 2	Relay
Pos Limit	In 3	Data Tables	Out 3	Pump1
Neg Limit	In 4		Out 4	Pump2
High Temp	In 5		Out 5	Solenoid 1
	In 6		Out 6	Solenoid 2
	In 7		Out 7	Solenoid 3

com

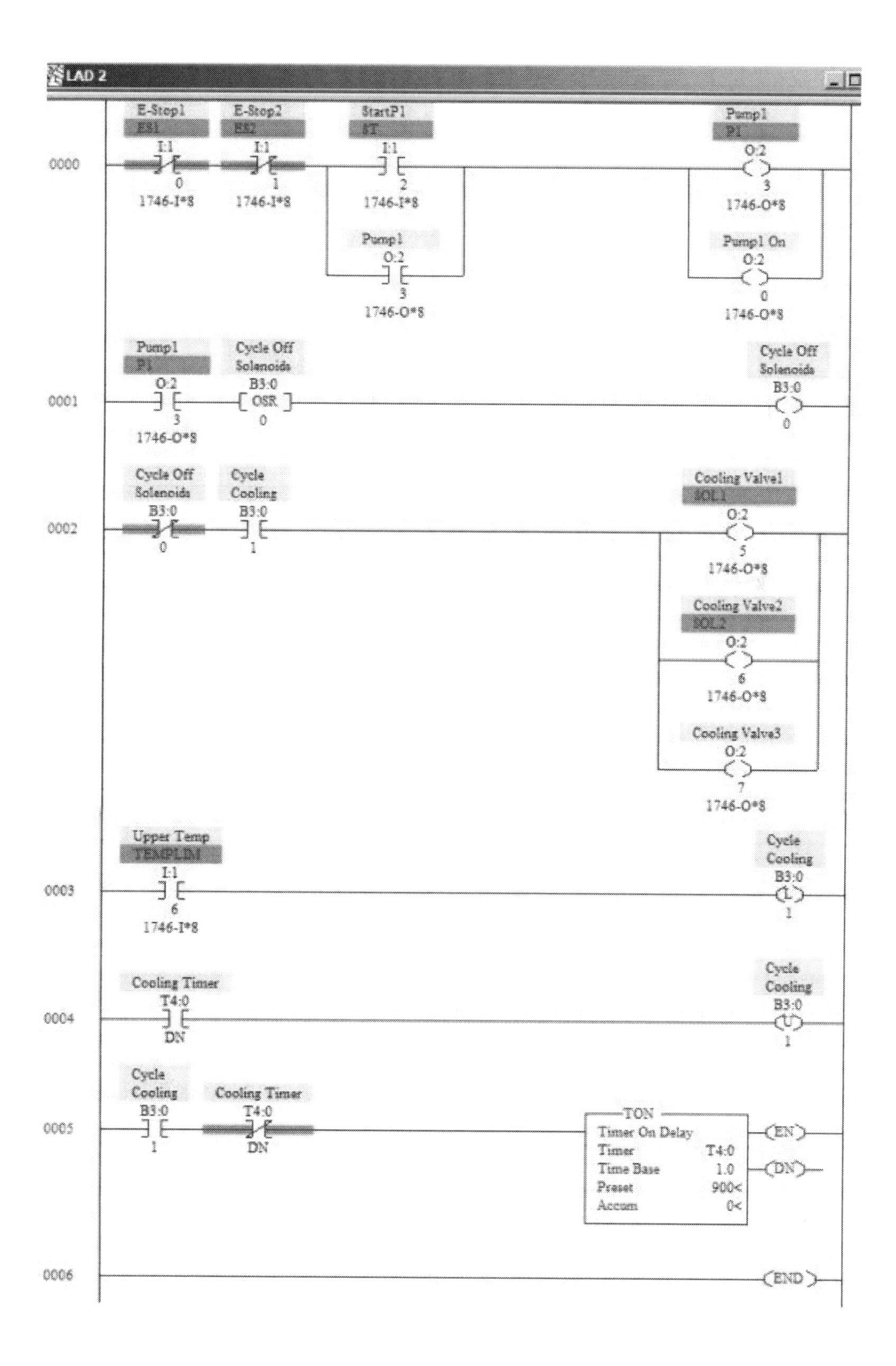

LAD 2
E-Stop1
ES1
I:1
0
1746-I*8
E-Stop2
ES2
I:1
1
1746-I*8
StartP1
ST
I:1
2
1746-I*8
Pump1
O:2
3
1746-O*8
Pump1
P1
O:2
3
1746-O*8
Pump1 On
O:2
0
1746-O*8
0000
Pump1
P1
O:2
3
1746-O*8
Cycle Off Solenoids
B3:0
OSR
0
Cycle Off Solenoids
B3:0
0
0001
Cycle Off Solenoids
B3:0
0
Cycle Cooling
B3:0
1
Cooling Valve1
SOL1
O:2
5
1746-O*8
Cooling Valve2
SOL2
O:2
6
1746-O*8
Cooling Valve3
O:2
7
1746-O*8
0002
Upper Temp
TEMPLIM
I:1
6
1746-I*8
Cycle Cooling
B3:0
L
1
0003
Cooling Timer
T4:0
DN
Cycle Cooling
B3:0
U
1
0004
Cycle Cooling
B3:0
1
Cooling Timer
T4:0
DN
TON
Timer On Delay
Timer T4:0
Time Base 1.0
Preset 900<
Accum 0<
EN
DN
0005
END
0006

Memory Usage:

I think it's a good time to mention some good programming practices that directly affect the use of available memory and the time it takes a PLC processor to complete a scan. If you remember I mentioned that a scan cycle can take from 5 to 50 milliseconds, so as you can see there is quite a range, and for most PLC applications you want the processor to update input and output tables with as little delay as possible. There are several good, sound programming "rules" that will help minimize the amount of memory used by your program.

One thing to note is that "relay" type instructions, such as an "XIC" or "XIO" take up a full word of memory. Remember also that we can quite literally use these instructions throughout a program as long as we address them properly. Rather than over-using that ability, we can utilize single bits of the "B3" data-file to program more efficiently.

Consider in the example below how the use of "bits" can provide efficiency in our programming and also save on memory use.

21 Words of memory usage!

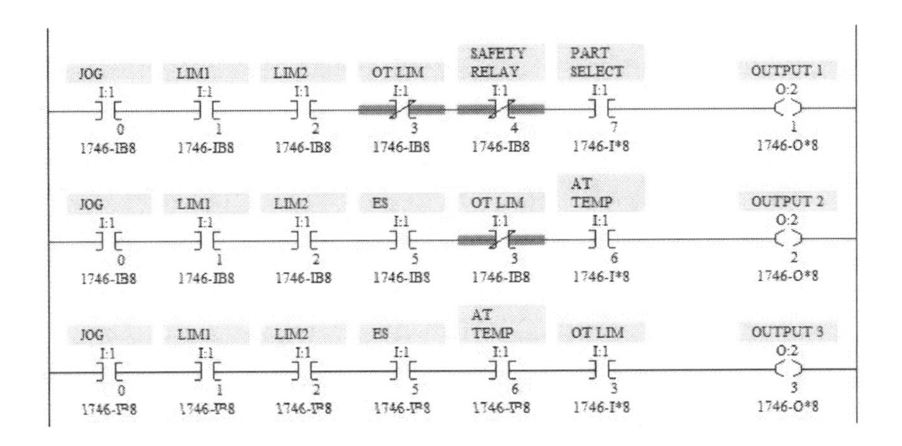

Note what these rungs have in common and replace multiple "relay" instructions with the addition of a single rung and use a designated bit as in the following example. This saves having repetitive use of the same three XIC instructions on multiple rungs and saves memory allocation as well!

19 Words of Memory Use!

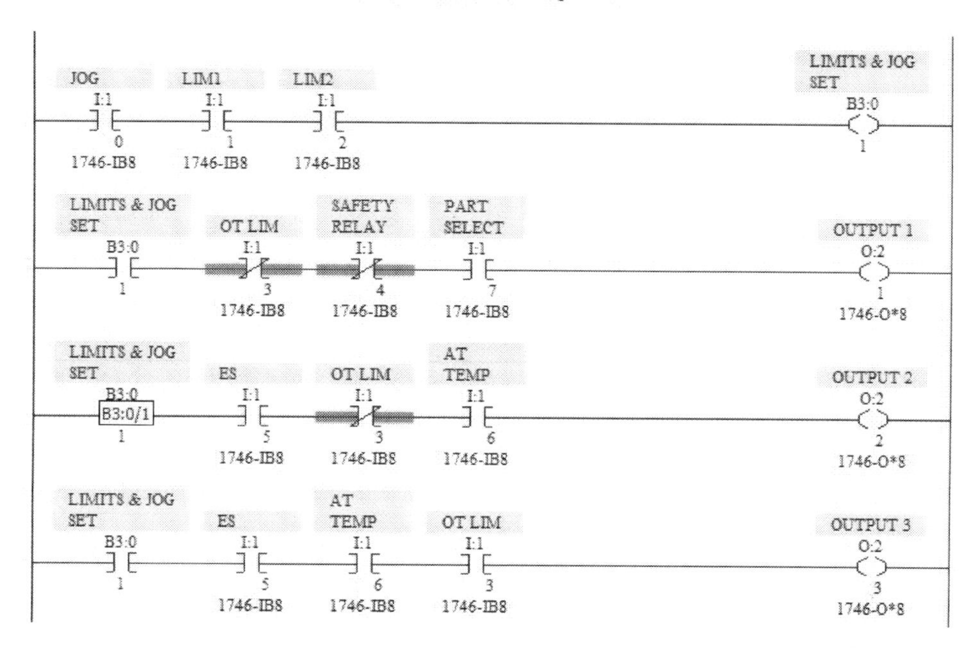

Another good programming practice is to avoid sloppy branching – as in a rung containing a multiple instruction "OR" scenario. Although a program will work as in the example below, each separate branch uses a word of memory. When the branching is done by extending branches already there, it uses less memory.

"Nested" Branch uses more memory than "Standard" branch!

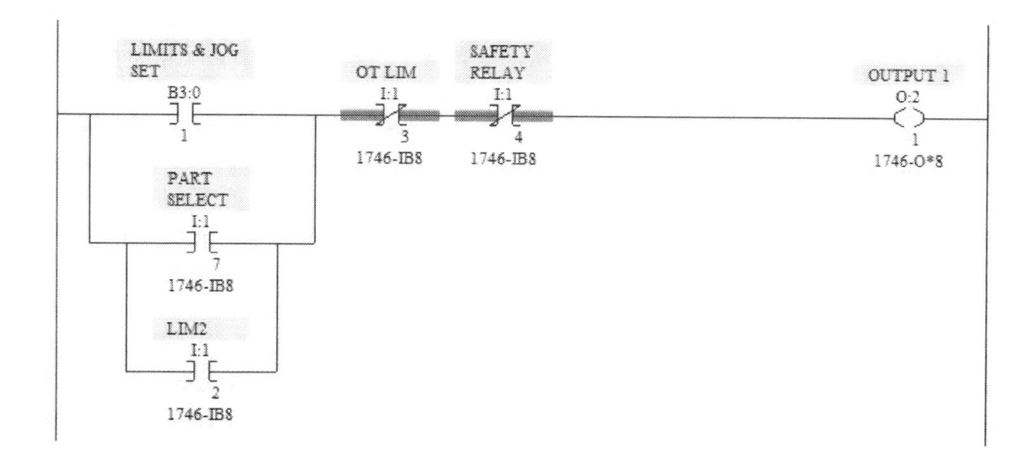

"Standard" Branching Example!

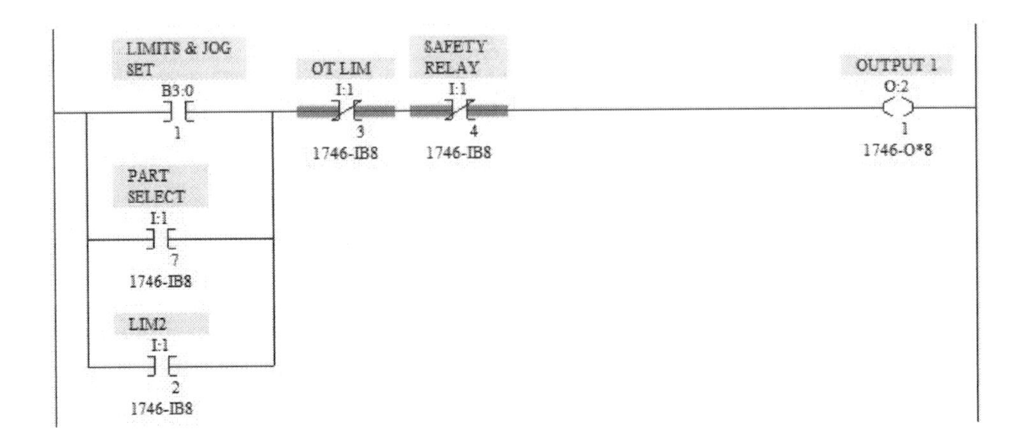

Perhaps one of the "most" important things to mention when discussing "good" programming practices, and the topic of the next chapter, is the whole area of "documentation". By providing descriptions, symbols and revision notes, you will help yourself or someone else have the ability to easily troubleshoot machinery or processes should something break down or if later alterations to the program need to be made.

Chapter 6: Program Documentation!

Just as electrical wiring schematics would be nearly impossible to follow without the words that tell what the components are and what they do, a PLC ladder logic program is nearly impossible to decipher without good documentation in the form of descriptions, symbols and comments. Considering the small samples of programming we have thus far looked into, it is easy to see the great value in using good descriptors.

This program could be difficult and time consuming when troubleshooting a problem!

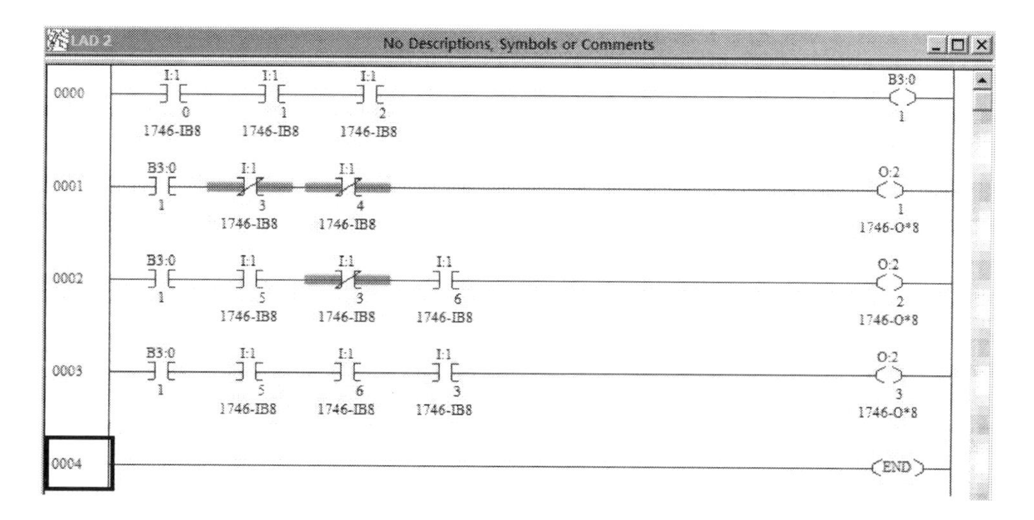

While the same program, with appropriate symbols and descriptions, would make the job much easier!

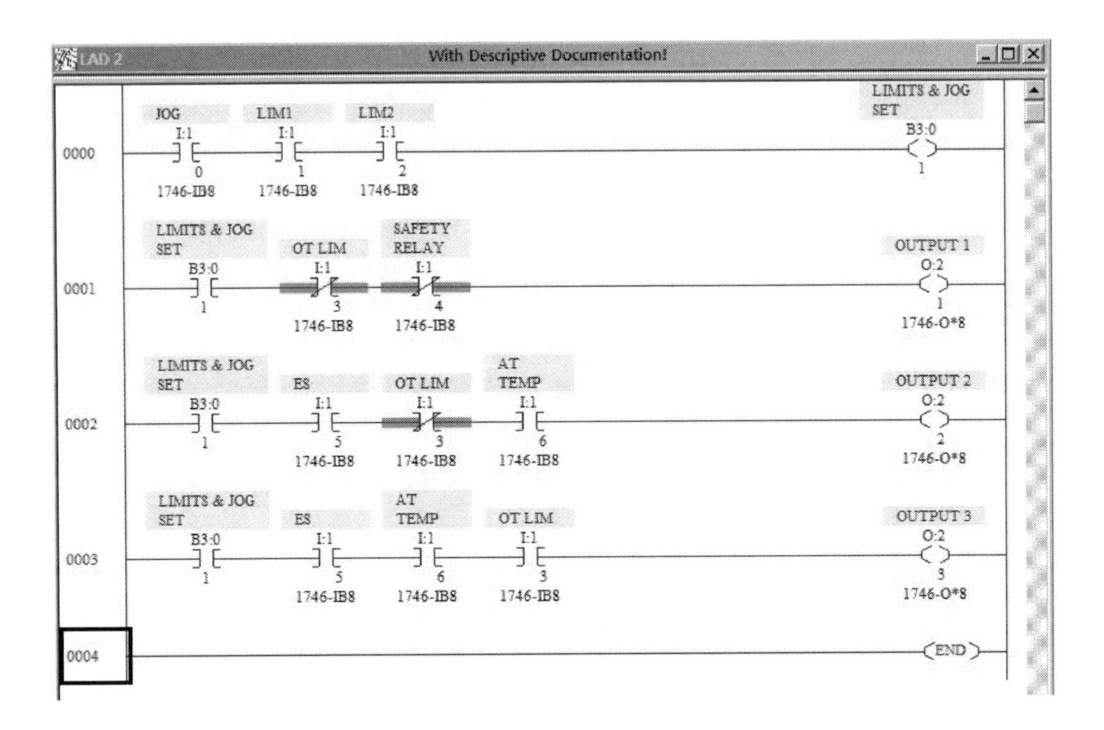

Address Symbols & Descriptions:

This type of documentation allows for easier *cross referencing* with electrical and process control schematics when locating or troubleshooting the field devices that make up the physical inputs to the PLC.

We have already seen some examples of one method of adding descriptions and symbols to your program - you simply enter descriptions and symbols to an instruction after entering the address. Another method is to go directly to the "Data File" where you can add descriptions and symbols and then link them to a specific address. They can then be utilized while programming actual instructions simply by entering the "symbol".

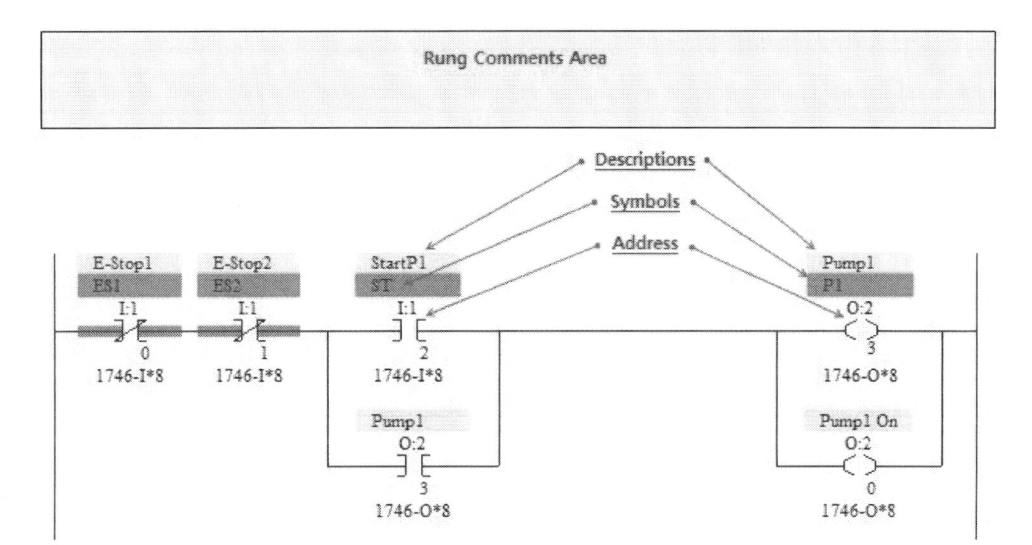

It would be somewhat easier, while programming, to simply enter "ST" or "P1" and allow the application software to fill in other appropriate information, such as the address, automatically. If you have taken time to build documentation into the data files (IE: the I1, O0, B3, N7, etc…) then all the descriptors you have should populate as you program each instruction and use either the symbol, address, or description with that instruction.

Here are the steps when using "method 2".

1. Open the appropriate "data-file" by "double-clicking" within the project tree.
2. Select the "bit", "terminal point", or "element".
3. Fill in the fields for "SYMBOL" and "DESCRIPTION".
4. Go to the next address bit/terminal/element and do the same.

Note: The examples below are shown with the ladder logic instructions in the background - just to give added information, but this 2nd method of filling out descriptor fields would normally be done before doing the actual programming. This is more of a configuration phase of the project and allows you to simply key-in the "symbols" as you are programming instructions.

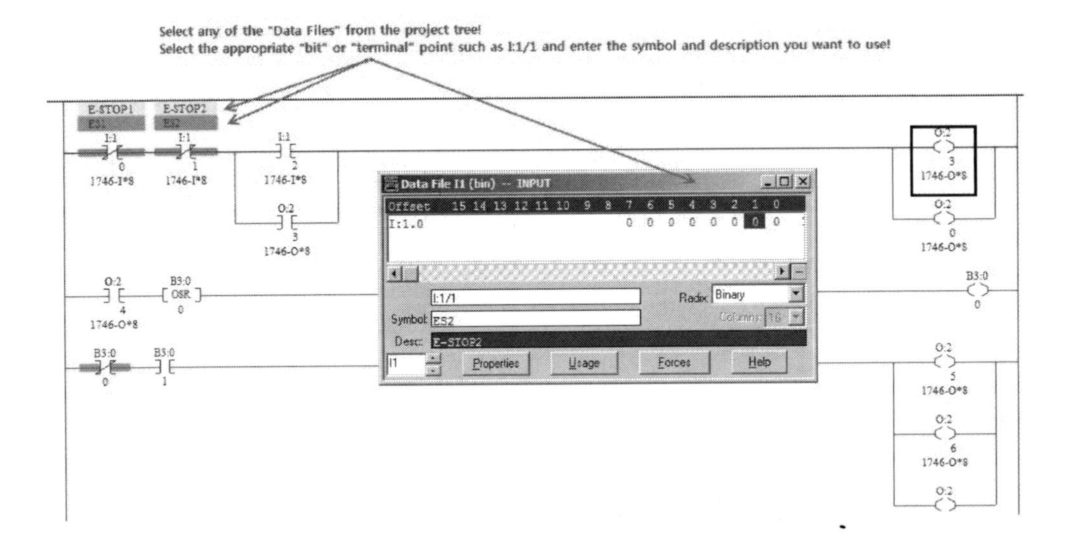

As you add an instruction, in this case an "XIC", they will appear with a "?" awaiting an address – you can key-in the "symbol" previously added to the data-file –in this case "P1", and the software will fill in the other fields. You can see the "P1" description and address added on the following examples.

As you add instructions they will appear with a "?" in the address field. Now you can just key in a symbol - such as "P1" and some selections that are now in the data file "I" will show up.
Select the appropriate one and all the fields will then be populated for that instruction!

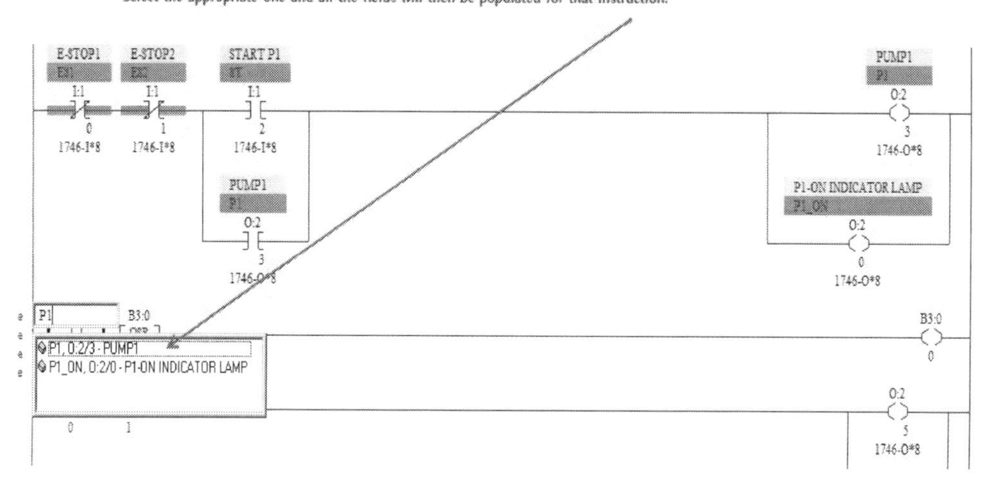

Descriptions and Symbols for other datafiles are accomplished the same way - as in this example of the "B3" datafile and the "T4" timer datafile that follows!

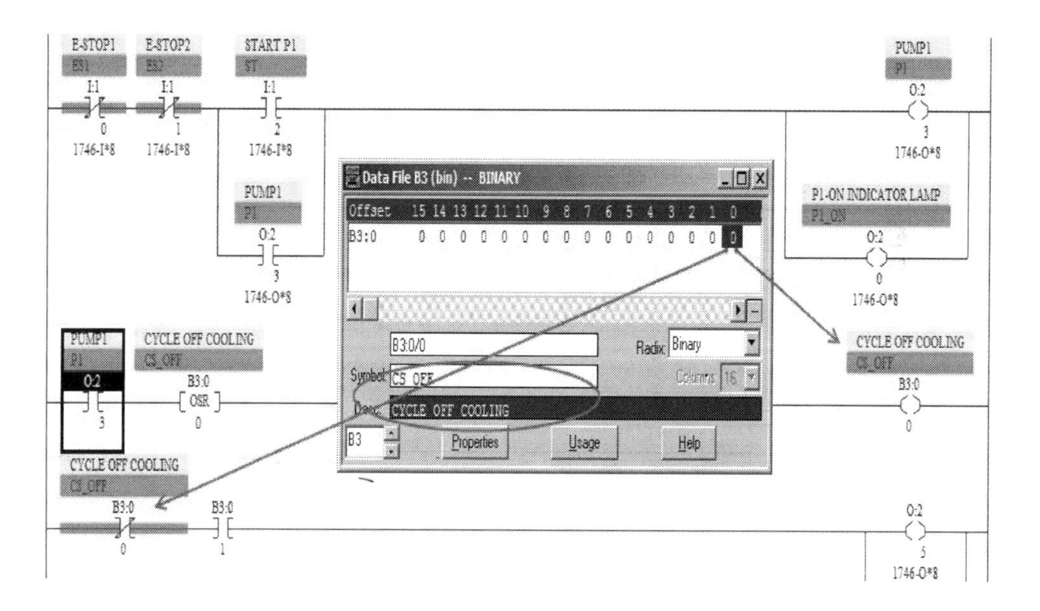

Add Descriptors to any specific "bit" or to the Timer itself!

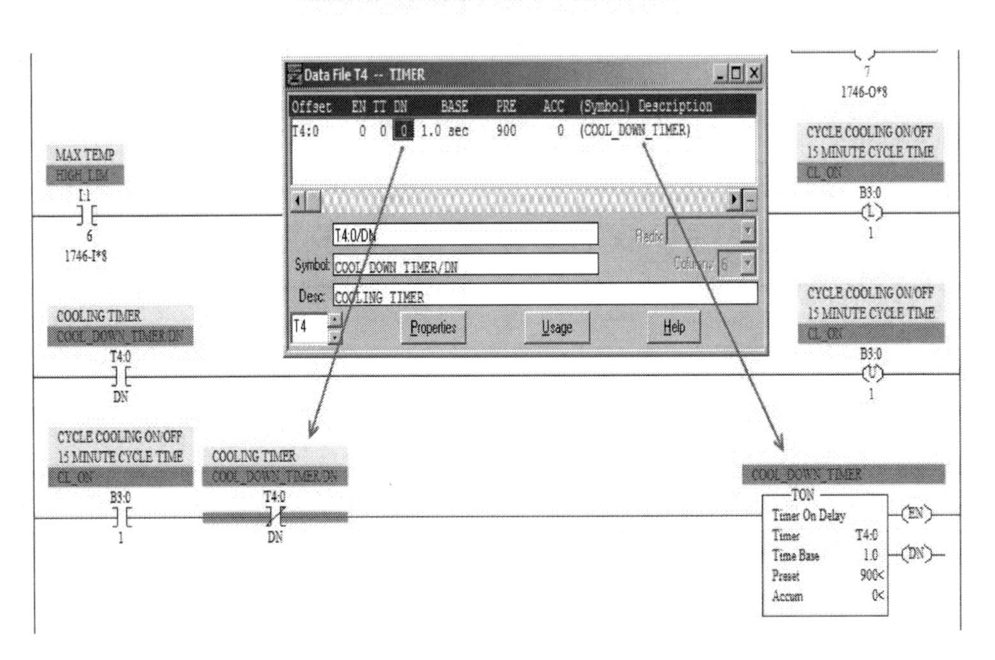

Rung Comments:

Helpful comments are easily added to a program by "clicking" on the left side rung number, it will highlight in red, and then doing a "right" click and selecting "*edit rung*". The following menu will be shown allowing you to enter your comments. Below are examples – before and after.

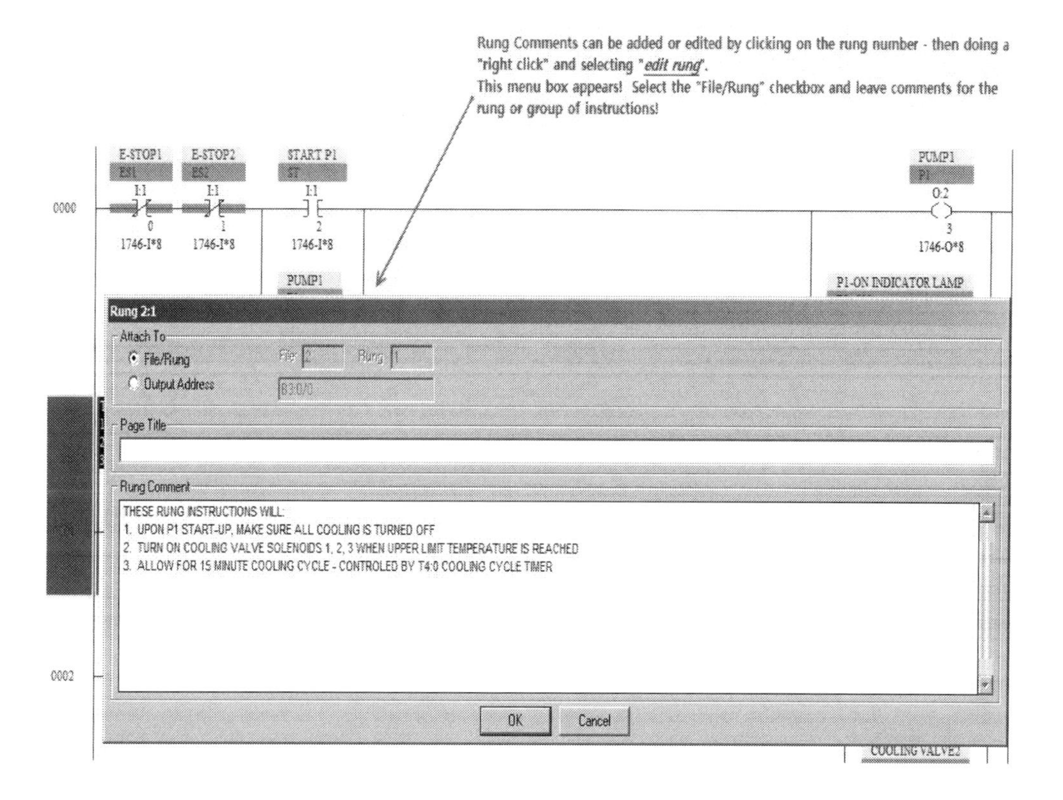

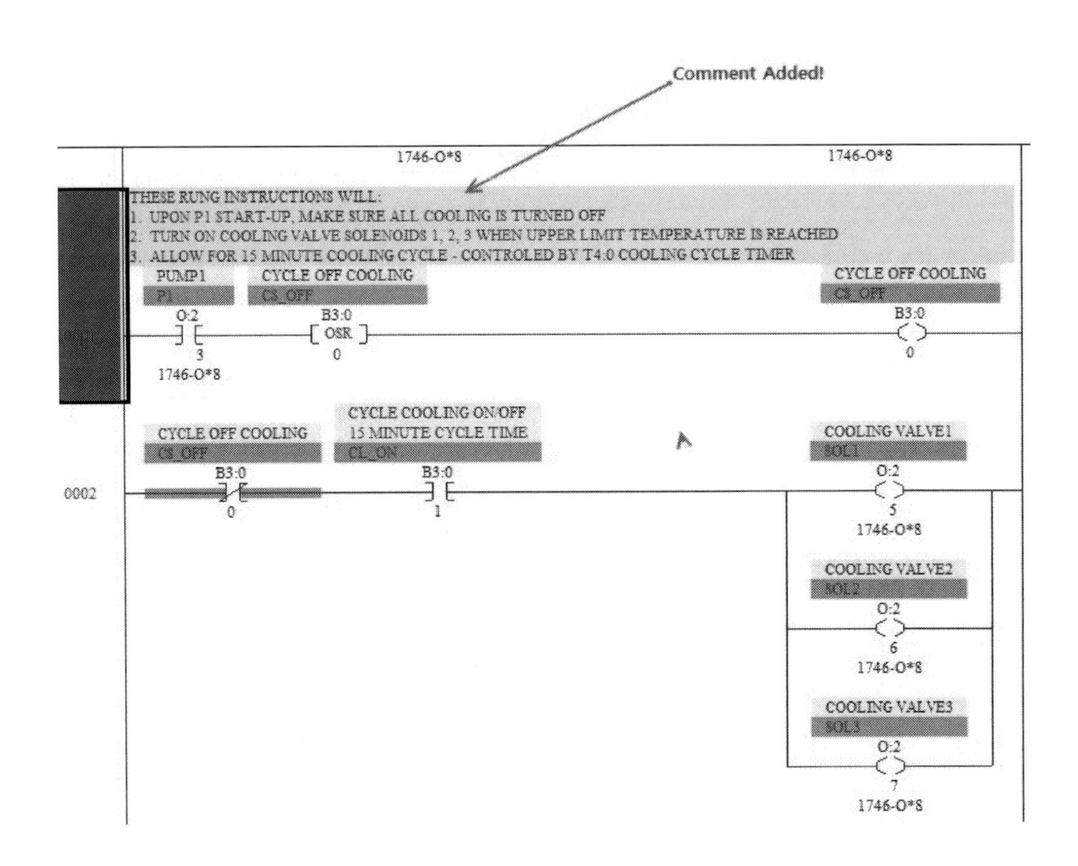

Comment Added!
1746-O*8
1746-O*8
THESE RUNG INSTRUCTIONS WILL:
1. UPON P1 START-UP, MAKE SURE ALL COOLING IS TURNED OFF
2. TURN ON COOLING VALVE SOLENOIDS 1, 2, 3 WHEN UPPER LIMIT TEMPERATURE IS REACHED
3. ALLOW FOR 15 MINUTE COOLING CYCLE - CONTROLED BY T4:0 COOLING CYCLE TIMER
PUMP1
P1
O:2
3
1746-O*8
CYCLE OFF COOLING
CS_OFF
B3:0
OSR
0
CYCLE OFF COOLING
CS_OFF
B3:0
0
0002
CYCLE OFF COOLING
CS_OFF
B3:0
0
CYCLE COOLING ON/OFF
15 MINUTE CYCLE TIME
CL_ON
B3:0
1
COOLING VALVE1
SOL1
O:2
5
1746-O*8
COOLING VALVE2
SOL2
O:2
6
1746-O*8
COOLING VALVE3
SOL3
O:2
7
1746-O*8

Verification:

Whenever the program is edited or new rung instructions are added, as in the case of building a new project, a vertical row of "e" will appear beside the rung. This is simply to indicate that since the program has been altered it needs to be compiled once again and checked for errors. You can do this by doing a "right click" on the red area and select the "verify rung" option. You can also select from the top menu tool bar the "**verify file**" or "**verify project**" buttons which will accomplish the same thing for the whole program.

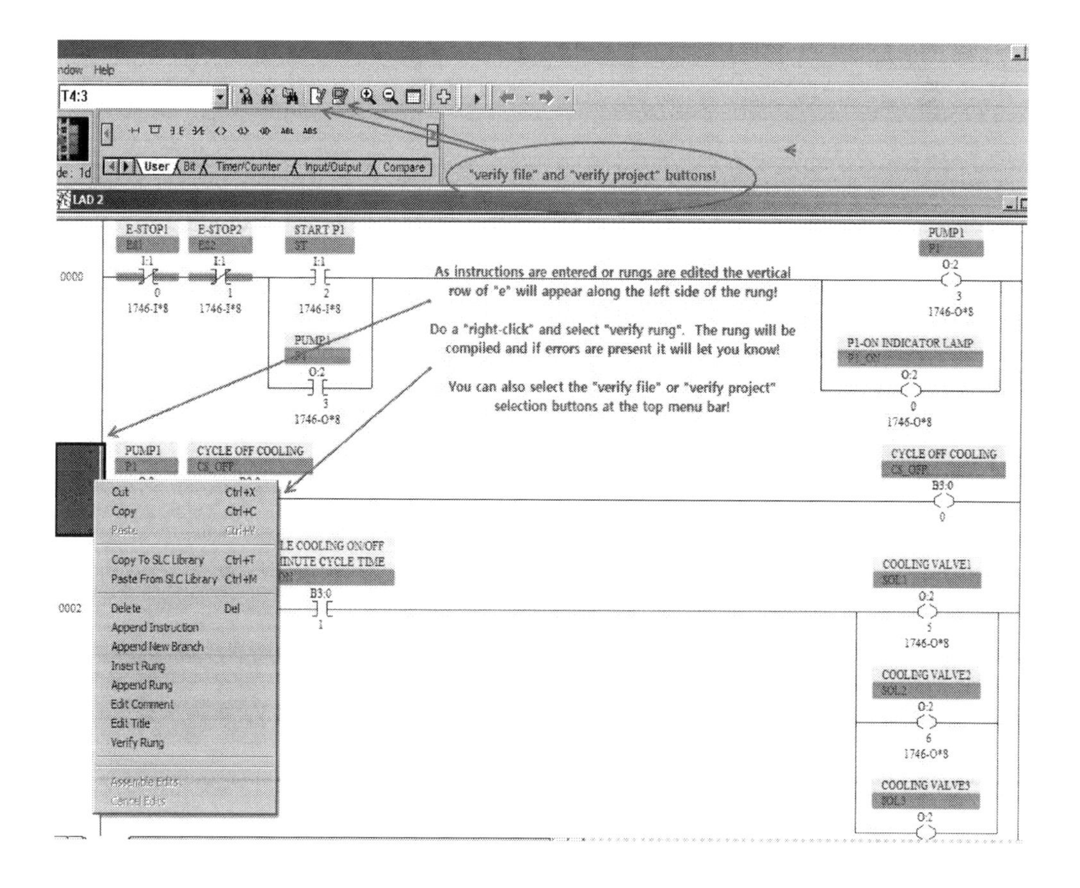

Remember to save the new or edited program by selecting the *"save"* or *"save as"* options from the file menu. Also note that while your program is ready to be "downloaded" into the PLC controller – only the ladder logic will be loaded into processor memory. All *comments, symbols and descriptions* continue to reside on the programming interface, such as your laptop. These can be "imported" back into your program on a different laptop or from a network if these "descriptor" files are saved on a network server.

Concluding Comments!

I have presented in this book, what I consider foundational concepts of ladder logic programming. Once again, the Allen Bradley family of PLC's are very good products, in my opinion, to learn programming and the basics of how PLC's provide machine control. Hopefully you have access to RSLogix500 software so you can develop and test your own programs. Most training seminars that teach PLC programming, provide laptops and PLC racks with I/O, and may schedule several days in the classroom. While this may be one of the best ways to learn these skills, it is not always an option to attend a three or four day class. The goal of this book is to convey these topics, as well as possible, from the written page – I hope I have accomplished that goal.

Some of the topics and objectives we have covered:

- PLC architecture and hardware components, including the power supply, processor or CPU, and several different types of I/O modules.

- How "input" and "output" modules function and the common wiring methods for machine control and for powering your PLC rack and I/O modules.

- Understanding the specifics of "*module-defined*" addressing and the differences between it and "*user-defined*" or "*tag based*" addressing.

- Basic "ladder logic" instructions including the "*question-asking*" type - the "XIC" and "XIO", and also the "*action-performing*" instructions such as the "OTE", OTL, OTU, OSR, and Timers (TON) & (TOF) have been presented.

- Boolean type expressions that are foundational to "*digital logic gates*" – the "AND", "OR", "NAND", "NOR", and "XOR" expressions accomplished with ladder programming.

- Several efficient and "*memory saving*" methods of programming, and what is going on during a "*scan cycle*" routine.

- Two methods on how to add documentation or "*descriptors*" to a ladder logic program, and the differences between "*symbols*", "*descriptions*" and "*comments*".

- How to *verify* your editing or newly written program so that it is ready to download into a PLC processor for use!

In the 2nd book of this series, "Advanced PLC Programming Concepts", more will be presented on "timers" and "counters", and how to *nest* these instructions to build unique timeframes for periodic alarms or maintenance cycles. Also presented will be details on the "register" type of instructions such as the MOV, COP, and the CLR, more on analog modules and analog scaling instructions, and also the "math" instructions that are integral to the RSLogix 500 software.

As we began this book, I mentioned how the first PLCs, back in the 60's were essentially built to replace the older "hard-wired" relay logic systems that were then commonly used. These types of control systems, which were on machine tools, automated welding equipment, in chemical process plants and a host of other environments were essentially composed of groups of relay contacts – both "normally open" and "normally closed", and different types of switches such as push buttons, limit, flow, and pressure switches, and timers. The basic RSLogix 500 instruction set covered in this book alone covers much, if not all, of the "functionality" of these older technologies.

If you are new to this type of programming but competent with working on electro-mechanical and electronic components, you may enjoy doing the modifications on some older equipment by integrating a SLC 500 or Micrologix PLC. It is one of the best ways to become comfortable and confident in your programming and troubleshooting skills.

To that end – I wish you all the best!

Other Books by Gary D. Anderson

Practical Guides for the Industrial Technician:

- *Motion Control for CNC & Robotics*
- *Variable Frequency Drives – Installation & Troubleshooting*
- *Industrial Network Basics*

RSLogix 500 Programming Series:

- *Basics Concepts of Ladder Logic Programming*
- *Advanced Programming Concepts*
- *Ladder Logic Diagnostics & Troubleshooting*
- *PID Programming Using RS Logix 500*
- *Program Flow Instructions Using RS Logix 500*

RSLogix 5000 & ControlLogix Controllers:

- *RSLogix 5000 - Understanding ControlLogix Basics*: this new book covers the many essential details and concepts that serve as building blocks for sucessful programming and troubleshooting within the ControlLogix platform.

As I've said before, I know you have many options when choosing from books and online resources that discuss these topics; so I thank you for selecting my book - I hope you feel you've benefited by doing so. As always, I welcome your comments and feedback. My goal is to present, in clear and concise language, relevant technical topics, and I have found feedback from my readers to be an invaluable resource. With that said, if you would like to contact me with questions or comments you can do so at the following email address:

Email: ganderson61@cox.net

Your reviews on Amazon are helpful and appreciated. If you've enjoyed what you've read and feel this book has provided a positive benefit to you, then please take a moment and write a short review.

Made in the USA
Lexington, KY
20 October 2017